HOLOCENE RECORDS OF NEBRASKA MAMMALS

Hugh H. Genoways
University of Nebraska State Museum
W436 Nebraska Hall
Lincoln, NE 68588-0514

Zea Books: Lincoln, Nebraska

2021

ISBN 978-1-60962-193-3 paperback
ISBN 978-1-60962-194-0 electronic book

doi:10.32873/unl.dc.zea.1277

Zea Books are published by the
University of Nebraska–Lincoln Libraries.

Digital ebook edition (pdf) available at
https://digitalcommons.unl.edu/zeabook/

Print edition can be ordered from
http://www.lulu.com/spotlight/unllib

Cover design by Ted Genoways
Cover map of Nebraska counties by Bruce Jones Design
Cover image of Bison skull via Shutterstock

University of Nebraska Lincoln

TABLE OF CONTENTS

ABSTRACT

A survey of the archeological and paleontological literature allowed a compilation of Holocene records of mammals in Nebraska. This survey identified Holocene records from 338 sites in 62 of the 93 Nebraska counties. These counties were located throughout state, but there was a concentration of sites in southwestern Nebraska where there were 27 fossil sites in Frontier County and 22 in Harlan County. Fossils sites were underrepresented in the Sand Hills region. Records of fossil mammals covered the entire Holocene period from 13,000 years ago until AD 1850. A minimum of 57 species (with eight additional species potentially present) representing six orders of mammals were represented in the compilation—four species of Lagomorpha, four species of Soricomorpha, 17 species of Carnivora (with three additional species potentially present), one species of Perissodactyla, six species of Artiodactyla, and 25 species of Rodentia (with five additional species potentially present). The remains of bison were found at 276 sites, which was more than for any other species in the state. Additional species that formed the main portion of the diet of Native Americans were the next most abundant in the fossil record—deer, pronghorn, and wapiti. That these food species dominated in the Holocene record was to be expected because fossils were recovered primarily from archeological sites.

INTRODUCTION

What is the Holocene? And what is not the Holocene? These would seem to be relatively straightforward questions, but in reality they are difficult to answer in an exact form. The Holocene is the most recent geological time period, following the end of the Wisconsonian glaciation at the end of the Pleistocene. The climate changes at the end of the Pleistocene were major and would not be expected to happen simultaneously in a state the size of Nebraska, which is 695 km [430 mi.] east to west and 340 km [210 mi.] north to south, with an elevation of 1654 m [5426 ft] in the southwestern Panhandle to 256 m [840 ft] in the southeastern corner of the state. Also local environmental conditions can slow or speed this transition from one geological period to another. Some archeological/paleontological sites may capture this transition, and examining their mammalian fauna can go a long way to defining the beginning of the Holocene. For the purposes of this research, I have defined the Holocene as that time period from the end of Pleistocene glaciation (13,000 years ago) until 1850. This was a time of ameliorating climate and the movement of mammal species in response to the retreating ice sheet. By mid-Holocene, mammals were moving into their modern distributional patterns and the modern biomes were developing.

When the Laurentide Ice Sheet retreated to the north as the Pleistocene was ending, it opened areas previously occupied by ice allowing plants and animals to shift their geographic ranges to the north in individualistic patterns (Davis and Shaw, 2001). Mayewski et al. (2004) studying Holocene global climates concluded that their variability was under the control of multiple factors, but their impact on various sites did not occur at the same time or with the same intensity. Mean July temperatures increased in North America by 3 to 4°C between 10,000 and 14,000 years ago, but the increase slowed in the early Holocene (Viau et al., 2006). Changes in annual precipitation were most obvious in the west and south of the northern Midwest where the precipitation decreased in the early Holocene and increased in the late Holocene (Bartlein et al.,

1984). This climatic variability in the late Pleistocene and Holocene has impacted mammals and their geographic ranges, which are still occurring as modern global climate changes continue.

The mammalian faunal changes concomitant with the end of the Pleistocene and the retreat of the continental glacier in North America has been well studied. Much of the Pleistocene megafauna was becoming extinct and the survivors were shifting their geographic ranges. The FAUNMAP Working Group (1996) concluded that mammals during the late Quaternary responded to the changing environmental conditions by a species reacting "in accordance with their individual tolerance limits, resulting in range shifts with varying rates, at different times, and in divergent directions." As the temperatures warmed at the end of the Pleistocene and into the Holocene, many mammals moved northward, not as a group but as individual species. Other species moved eastward and others westward, whereas some species remained in their existing geographic ranges. Lyons et al. (2010; Lyons, 2003) found that the size of these geographic range shifts was related to the body size, lifespan, and the topography the species encountered. Larger-bodied and longer-living species tended to expand their geographic ranges more than those species that were small bodied, with short lifespans.

Although the Holocene climatic changes were not as radical as those at the end of the Pleistocene, there were shifting climatic conditions, which have resulted in the recognition of at least nine post-glacial episodes (Wendland, 1978; Bartlein et al., 1984; Graham et al., 1987; Bozell, 1995): *Pre-Boreal* (10,000-9300 years ago)—in the east, climate zones were moving to the north and in the Great Plains, the grasslands were extending to the east as precipitation decreased and summer temperatures increased; *Boreal* (9300-8490 years ago)—climate borders still moving north and eastward as modern biomes were becoming organized, with the Laurentide glacier in the Great Lakes area; *Atlantic* (8490-5060 years ago)—maximum aridity and temperatures for the Holocene, with the conifer-hardwood forests nearing modern positions and the ice sheet centered over Hudson Bay; *Sub-Boreal* (5060-2760 years ago)—Laurentide ice wasting and by 4000 years ago the ecotones were in modern positions; *Sub-Atlantic* (2760-1680 years ago)—deteriorating climate north of 40° N and cooler than Sub-Boreal; *Scandic* (1680-1260 years ago)—a warming and drying trend; *Neo-Atlantic* (1260-850 years ago)—somewhat more moist in summer, approaching modern moisture scheme; *Pacific* (850-400 years ago)—drying climate; *Neo-Boreal* (400-100 years ago)—Little Ice Age, with cooler and moister climate. These climatic episodes applied broadly across North America, but did vary in timing and impact depending on the location under consideration.

Although the objective of this study was to examine both the archeological and paleontological records to gain an understanding of Holocene mammals in Nebraska, the reports were heavily weighted toward archeological data, which brings its own set of biases. Semken (1983) discussed this bias in the mammalian Holocene record toward archeological over paleontological reports. Because the Holocene mammalian fauna was basically modern in nature, paleontologists have tended not to prepare publications on these studies, whereas archeologists' work is concentrated in the Holocene and they have prepared numerous publications and reports on their studies.

The goal of the present study was to compile and analyze data for Holocene mammals in Nebraska to gain an understanding of specifically how this fauna changed through the era. We can determine what species have appeared or disappeared from the state during the Holocene. The compiled records document changes in the geographic distribution of individual species and may allow hypotheses to be developed as to what forces have caused

Table 1.—List of mammalian species identified from a late Pleistocene deposit at the North Cove site (25HN164), Harlan County Lake, Harlan County, Nebraska (Adair and Brown, 1987; Adair 1989).

Extinct species

Mammut americanum—American mastodon
Navahoceros fricki—American mountain deer
Bison occidentalis—extinct bison
Martes nobilis—noble martin
Urocitellus kimballensis—Kimball ground squirrels

Species living in southwestern Nebraska

Blarina brevicauda—northern short-tailed shrew
Mustela frenata—long-tailed weasel
Ictidomys tridecemlineatus—thirteen-lined ground squirrel
Cynomys sp.—prairie dog
Geomys sp.—pocket gopher
Microtus ochrogaster—prairie vole
Microtus pennsylvanicus—meadow vole
Ondatra zibethicus—muskrat
Peromyscus cf. *leucopus*—white-footed mouse
Peromyscus maniculatus—deer mouse

Species living elsewhere in Nebraska

Sorex cinereus—masked shrew
Tamias minimus—least chipmunk
Thomomys talpoides—northern pocket gopher
Zapus hudsonius—meadow jumping mouse

Species living outside of Nebraska

Lepus americanus—snowshoe hare
Sorex arcticus—Arctic shrew
Sorex cf. *palustris*—American water shrew
cf. *Martes pennanti*—fisher
Tamiasciurus hudsonicus—American red squirrel
Marmota cf. *flaviventris*—yellow-bellied marmot
Zapus princeps—western jumping mouse
Myodes gapperi—southern red-backed vole
Microtus montanus—montane vole
Microtus cf. *richardsoni*—North American water vole
Microtus xanthognathus—Taiga vole
Phenacomys intermedius—western heather vole
Synaptomys borealis—northern bog lemming

these changes. This examination begins with a look at the Pleistocene to Holocene transition as it occurred in Nebraska.

Table 1 presents the mammalian species identified from a latest Pleistocene deposit at the North Cove site (25HN164), Harlan County Lake, Harlan County, Nebraska (Adair and Brown, 1987; Adair 1989). The deposits were 10,000 to 14,700 years old. The flora of southwestern Nebraska during this time period was "an open, spruce-deciduous parkland," with such trees as Engelmann and white spruce, aspen, oak, hornbeam, birch, and alder. Other plants included sagebrush, giant ragweed, sunflower, goosefoot, pigweed, switchgrass, blue grama, and buffalo grass.

This ice age fauna is an excellent example of the composition of these faunas on the Great Plains and will illustrate the changes that occur at the beginning of the Holocene. There are five extinct species of mammals in the Pleistocene fauna, whereas there are no extinct species represented in the Holocene faunas in Nebraska unless they have occurred in the last 100 years. Among the extinct species are two ice age megafauna—the American mastodon (*Mammut americanum*) and Pleistocene bison (*B. occidentalis*), which was much larger than the modern bison. The largest group is the extant species that occur in the boreal forests of northern North America and/or in the montane areas of the western United States. "All of the pollen and spore taxa identified [at the North Cove site] are present today in the boreal forests and plains of Canada."

There are four species of mammals that have retreated from southwestern Nebraska, but have not gone nearly as far away as the previous 13 species and remain as part of the Nebraska fauna. *Zapus* and *Sorex* are widespread in the state but tend to select the most mesic habitats. The other two species occupy very limited areas of the northwestern Panhandle (*Tamias* and *Thomomys*) and the southwestern Panhandle (*Thomomys*). The remaining 10 species are still present in southwestern Nebraska. The compressing of widespread faunas is one of the characteristics of Pleistocene mammalian faunas. As the Pleistocene ended, the changing climates and floras resulted in decompression of faunas with the species moving in many directions. Many, but certainly not all, were moving northward as the continental glaciers retreated. Other species simply remained in place during the Holocene. The North Cove mammalian fauna clearly demonstrates these movement patterns. From this point, we begin the survey of the Holocene mammalian fauna of Nebraska.

METHODS AND MATERIALS

Holocene records of mammals in Nebraska are based on a survey of the archeological and paleontological literature. The literature search for a number of species began with data from FAUNMAP. The Holocene literature, particularly reports on archeological sites, is extremely scattered and presents significant challenges in completing a thorough survey. These reports range from books published by major university and commercial presses, to reports kept in the files of individual investigators or historical organizations. Only one or two copies of some reports have ever existed. The internet was a vital resource in identifying the sources of many survey reports and more of the survey reports are becoming available on the world wide web. I hope that my efforts have found and evaluated as much as 90% of this relevant literature. The literature survey was finished approximately at the end of 2010. The publication of this material has been delayed by the author's health and other personal issues.

The identification of remains is based on the original reports. I have included only those records when the investigators definitely identified the remains as those of a mammal. There are some challenges in interpreting the significance of some archeology records. These records are associated primarily with human hunting activity, which probably resulted in at least short distance transportation of the remains. Skeletal remains that were worked into utilitarian and ceremonial objects could be subject to even longer distance transportation by humans. Early archeological workers did not preserve animal bones associated with human artifacts. When bones were preserved, the emphasis was on the bones of larger mammals that constituted the diets of the human occupants of the site. However, beginning in the 1960's archeologists' efforts extended to fine screening and floatation methods that allowed the recovery of small bones and plant materials, which greatly enhanced the data available from the sites. The archeological sites that have been studied are not spread uniformly over the state. Many studies were undertaken as the result of highway construction projects and the building of dams and creation of the associated lakes. For example, the Sand Hills have not been as intensely studied as other areas in the state, whereas sites in Frontier and Red Willow cos. have been extensively investigated.

Species Accounts

The 72 following accounts cover individual species or are generic accounts for material that could not be identified to the species level. The list of accounts follows the order of species as listed in Wilson and Reeder (2005; Genoways et al., 2008). Each account includes site records arranged by counties in alphabetical order and within counties the material is listed in numerical order based on site numbers. Each account also includes my remarks based on observation of the site records and the reports that accompanied them.

ORDER LAGOMORPHA

Sylvilagus audubonii—desert cottontail

Site records (6).—Frontier Co.: 25FT22, Sec. 11, T5N, R26W (Turnmire, 1996: 189-191); 25FT54, Red Willow Creek, 5 mi. upstream from Red Willow Dam (Grange, 1980: 170; Mick, 1983: 168); 25FT58, Red Willow Creek, ≈4.5 mi. upstream from Red Willow Dam (Grange, 1980: 171); 25FT80, Red Willow Creek, 4.5 mi. upstream from Red Willow Dam (Grange, 1980: 171; Mick, 1983: 168). Red Willow Co.: 25RW22, 3 mi. N, 6.25 mi. W Indianola (Grange 1980: 172; Mick, 1983: 168); Doyle Site, 25RW28, 9.5 mi. N, 0.75 mi. W McCook (Grange, 1980: 172).

Remarks.—Modern populations of the desert cottontail occur in the western third of Nebraska, including the eastern-most recent records in extreme southwestern Nebraska from Chase and Dundy cos. (Jones, 1964). The six Holocene records are from Frontier and Red Willow cos., which are about 100 km [60 mi.] to the east of the boundary of modern geographic range. The remains from all four sites in Frontier Co. and one from Red Willow Co. date from between 500 to 1000 years ago. The Doyle Site in Red Willow Co. dates from 1270 to 2050 years ago.

Sylvilagus floridanus—eastern cottontail

Site records (28).—Burt Co.: Logan Creek Site A, D, 25BT3, Logan Creek, SW Oakland (Snyder and Bozell, 1983: 16; Widga, 2006: 67-68). Butler Co.: Palmer Johnson Site, 25BU37, east of Ulysses (Souders, 1994: 53; Bozell and Koch, n.d.). Cass Co.: Walker Gilmore, 25CC28, 6 mi. SE Murray (Haas, 1983: 251; Graham et al., 1987: 284). Dixon Co.: Annie's Site, 25DX30, in vicinity of 2 mi. N, 3 mi. E Newcastle (Manz and Blakeslee, 1988: 159-160). Douglas Co.: Parker Site, 25DO2, on the bluffs north of Ponca Creek and west of the Missouri River, in extreme northeastern Douglas Co. (Ewing, 2000: 4). Frontier Co.: 25FT13, 16, 17, 70, vicinity of Medicine Creek Dam (Mick, 1983: 173, 183, 189, 209: Graham et al., 1987: 284, 286; Kivett and Metcalf, 1997: 213); 25FT20, 30, near point where Lime Creek originally entered Medicine Creek (Mick, 1983: 201; Graham et al., 1987: 286; Kivett and Metcalf, 1997: 213); 25FT22, ≈ 4 km upstream on Medicine Creek from Medicine Creek Dam (Turnmire, 1996: 189-191; Kivett and Metcalf, 1997: 213); Mowry Bluff Site, 25FT35, 6 mi. NW Cambridge (Falk, 1969b: 47; Mick, 1983: 204; Graham et al., 1987: 286); 25FT36, ≈ 4 km upstream on Medicine Creek from Medicine Creek Dam (Mick, 1983: 206; Kivett and Metcalf, 1997: 213); 25FT39, ≈ 5 km upstream on Medicine Creek from Medicine Creek Dam (Mick, 1983: 209; Graham et al., 1987: 286; Kivett and Metcalf, 1997: 213). Garden Co.: Ash Hollow Cave Site, 25GD2, 3 mi. SE Lewellen (Champe, 1946: 43-44). Hall Co.: Hulme Site, 25HL28, 16 mi. W Grand Island (Bozell, 1991: 233; Souders, 1994: 53). Howard Co.: Schmidt Site, 25HW301, along North Loup River, near Elba (Mick, 1983: 155; Graham et al., 1987: 284). Nance Co.: 25NC8, 10, 12, vicinity of Fullerton (Mick, 1983: 159, 162, 165; Graham et al., 1987: 284). Richardson Co.: Rulo Southeast, 25RH69, 0.4 km S Rulo (Nelson, 2006: 165); Rulo Southeast, 25RH70, 0.6 km S Rulo (Nelson, 2006: 170, 173). Sarpy Co.: Lucien Fontenelle's Post, 25SY26, Bellevue (Bozell et al., 1990: 29); Patterson Site, 25SY31, ≈ 7 mi. S Gretna (Bozell and Ludwickson, 1999: 83). Sherman Co.: Bill Packer Site, 25SM9, along Davis Creek in extreme northeastern corner of county (Graham et al., 1987: 284; Bozell and Rogers, 1989: 27; Souders, 1994: 53). Webster Co.: Shipman Site, 25WT7, south side of Republican River, between Red Cloud and Guide Rock (Mick, 1983: 153; Graham et al., 1987: 286).

Remarks.—The eastern cottontail currently is statewide in its distribution in Nebraska. It is sympatric with the desert cottontail in the western third of the state where it occupies more mesic habitats in low-lying areas along streams and rivers as opposed to the drier upland habitats used by S. *audubonii* (Jones, 1964; Genoways et al., 2008b). The eastern cottontail is more common in the Holocene record than the desert cottontail being reported from 28 sites. These sites are scattered almost throughout the state with four counties from along the eastern boundary of the state (Burt, Cass, Douglas, and Sarpy), Richardson Co. in the southeast, Dixon Co. in the northeast, and a cluster of sites in five counties in the east-central portion of Nebraska (Butler, Hall, Howard, Nance, and Sherman cos.). The record from the Ash Hollow Cave Site in Garden Co. is the farthest west Holocene locality so that in large areas of the Panhandle and Sand Hills there are no Holocene records. The remaining Holocene records are from one site in Webster Co. along the southern boundary of the state and nine sites in Frontier Co. in the southwestern corner of the state. Frontier is the only county where there are Holocene records of both S. *audubonii* and S. *floridanus*, but the two species share only one specific site, 25FT22. The oldest Holocene eastern cottontail material is from the Logan Creek Site, level D, Burt Co., which dates from approximately 7000 years ago.

***Sylvilagus* sp.—cottontail**

Site records (27).—Buffalo Co.: 25BF128, Sec. 11, T9N, R16W (Ludwickson, 1978b: 34); 25BF173, 3.5 mi. E Miller (Ludwickson, 1978b: 35). Cass Co.: Theodore Davis Site, 25CC17, 1.5 mi. E Weeping Water (Gradwohl, 1969: 143). Cedar Co.: Ferber Site, 25CD10, near Bow Valley Mill (Ludwickson et al., 1981: 22-23). Cherry Co.: 25CE255, 3 mi. S, 2 mi. W Sparks (Pepperl and Falk, 1983: B83); 25CE312, 3.5 mi. S, 1 mi. W Sparks (Pepperl and Falk, 1983: B89). Dakota Co.: Big Village Site, 25DK5, 1 mi. NE Homer (O'Shea and Ludwickson, 1992: 331). Dawes Co.: 25DW59, 0.7 mi. W Crawford (Bozell and Ludwickson, 1988: 86). Hooker Co.: Humphrey Site, 25HO21, Middle Loup River, about 5 mi. E Mullen (Gunnerson, 1960: 204). Franklin Co.: Lost Creek Site, 25FR6, Lost Creek, 2.25 mi. S, 1.5 mi. E Bloomington (Strong, 1935: 100). Frontier Co.: 25FT18, near point where Lime Creek originally entered Medicine Creek (Kivett and Metcalf, 1997: 212); 25FT22, Sec. 11, T5N, R26W

(Turnmire, 1996: 189-191); 25FT31, Red Willow Creek, adjacent to Red Willow Dam (Grange, 1980: 166); Lime Creek Site, 25FT41, north side of Lime Creek Valley, 1 mi. from original junction of Lime and Medicine creeks (Graham et al., 1987: 283); Allen Site, 25FT50, ≈ 3 mi. upstream from Medicine Creek Dam (Bamforth, 2002: 65; Hudson, 2007: 195). Garden Co.: Ash Hollow Cave Site, 25GD2, 3 mi. SE Lewellen (Champe, 1946: 43-44). Gosper Co.: Wallace Site, 25GO2, Plum Creek, 5.5 mi. N, 9.5 mi. E Elwood (Winfrey, 1991: 82). Harlan Co.: 25HN12, 6 mi. S Alma (Adair and Brown, 1987: 100, 573); 25HN36, 2 mi. S Alma (Adair and Brown, 1987: 153, 581-582); 25HN40, 0.8 mi. S, 2 mi. E Alma (Adair and Brown, 1987: 194, 590). Hooker Co.: Kelso Site, 25HO23, Middle Loup River, about 5 mi. E Mullen (Kivett, 1952: 39-40). Morrill Co.: Greenwood Site, old Greenwood Stage Station, Keenan Ranch, ≈ 9 mi. S, 8 mi. E Redington (Renaud, 1933: 14). Nance Co.: Palmer Locality [in part], 25NC29, 5.5 mi. S, 19 mi. W Fullerton (Meadow and Peterson, 2001: 156). Otoe Co.: 25OT24, Sec. 26, T8N, R14E (Kaldahl, 1993: 106-112). Scotts Bluff Co.: 25SF76, Scotts Bluff National Monument, 0.7 mi. N, 0.1 mi. W Visitor Center (Bozell, 1994b: 52, 63). Sioux Co.: 25SX157, Agate Fossil Beds National Monument, 19.25 mi. S, 5.75 mi. E Harrison (Bozell, 1994a: 47, 62). Valley Co.: Mira Creek Site, 25VY1, 6 mi. W, 0.75 mi. N North Loup (Hill and Kivett, 1940: 183).

Remarks.—The cottontails in Nebraska are identified based on external and cranial characteristics, but are impossible to separate based on broken cranial material and post-cranial skeletons, which has resulted in some authors trying to use DNA to identify their rabbit subfossils (Yang et al., 2005). Based on our current knowledge, we can presume that records from the eastern half of Nebraska are attributable to *S. floridanus*, but in the western half of the state either species should be expected.

These small rabbits were probably taken as a food item at times so the skeletal remains should originate from the local area.

***Lepus californicus*—black-tailed jackrabbit**

Site records (5).—Brown Co.: McIntosh Site, 25BW15, Enders Lake, ≈12 mi. S, 12.5 mi. W Ainsworth (Koch, 1995: 43; 2004: 116). Frontier Co.: 25FT39, ≈ 5 km upstream on Medicine Creek from Medicine Creek Dam (Kivett and Metcalf, 1997: 213); 25FT80, Red Willow Creek, 4.5 mi. upstream from Red Willow Dam (Grange, 1980: 171; Mick, 1983: 168). Red Willow Co.: 25RW22, 3 mi. N, 6.25 mi. W Indianola (Grange 1980: 172; Mick, 1983: 168); Doyle Site, 25RW28, 9.5 mi. N, 0.75 mi. W McCook (Grange, 1980: 172).

Remarks.—Black-tailed jackrabbits are found throughout the state of Nebraska, but as Jones (1964) stated: "Possibly *L. californicus* did not occur at all in Nebraska 100 years ago." He based his hypothesis on the fact that early collectors took only white-tailed jackrabbits along the Platte River where *L. californicus* is now the only jackrabbit present. Nebraska is near the northeastern limit of the geographic range of the black-tailed jackrabbit because the species goes no further north of the Nebraska border than a third of the way into southern South Dakota (Hall, 1981).

There are five Holocene records from Nebraska, which would seem to argue for the fact that the species has been present in the state far longer than the last 100 years. Four of the sites are in Frontier and Red Willow cos. in southwestern Nebraska and the fifth site, interestingly, is in Brown Co. in northern Nebraska near the northern distributional limit of modern *L. californicus*. Material from the Doyle Site in Red Willow Co. is the oldest in our sample being from between 1300 and 2000 years ago. The remaining sites fall into the age range of AD 1200 to 1500. The northern-most site in Brown Co. is from AD 1200 to 1450. This does leave a 400 year gap in the record from 1450 to 1850 when the geographic range of the black-tailed jackrabbits could have contracted to the south of the state before it began to re-expand beginning around 1900. However, this does seem like a lot of movement not seen in other species occurring in Nebraska, but this is certainly a mobile species.

Lepus townsendii—white-tailed jackrabbit

Site records (13).—Frontier Co.: 25FT13, 16, 17, 70, vicinity of Medicine Creek Dam (Mick, 1983: 173, 184, 189, 209: Graham et al., 1987: 284, 286; Kivett and Metcalf, 1997: 213); 25FT18, 30, near point where Lime Creek originally entered Medicine Creek (Mick, 1983: 201; Graham et al., 1987: 286; Kivett and Metcalf, 1997: 212-213); 25FT22, Sec. 11, T5N, R26W (Turnmire, 1996: 189-191); Mowry Bluff Site, 25FT35, 6 mi. NW Cambridge (Falk, 1969a: 47; Mick, 1983: 204; Graham et al., 1987: 286); 25FT36, ≈ 4 km upstream on Medicine Creek from Medicine Creek Dam (Mick, 1983: 206; Kivett and Metcalf, 1997: 213); 25FT39, ≈ 5 km upstream on Medicine Creek from Medicine Creek Dam (Mick, 1983: 209; Graham et al., 1987: 286). Harlan Co.: 25HN65, 1.25 mi. W Alma (Moore, 1988: 145). Howard Co.: Schmidt Site, 25HW301, along North Loup River, near Elba (Mick, 1983: 155). Loup Co.: Gracie Creek Site, 25LP16, 0.5 mi. upstream on Gracie Creek from its confluence with the Calamus River (Falk and Pepperl, 1981: A1 151).

Remarks.—Jones (1964) believed that the white-tailed jackrabbit was statewide in distribution with the exception of the southeastern-most counties prior to European settlement of Nebraska. By the time that Jones (1964; Genoways et al., 2008b) was preparing his monograph, he believed that the white-tailed jackrabbit was absent south of the Platte River in southeastern and south-central Nebraska. Jones (1964) hypothesized that the breaking of prairies and other natural vegetation and its conversion to agricultural purposes favored the smaller black-tailed jackrabbit and resulted in the retreat of the white-tailed jackrabbit northward. Benedict et al. (2000) believed the species may have retreated even further north since Jones' documentation.

The Holocene records from Howard and Loup counties are within the geographic range proposed for the species even as agriculture was being introduced. These sites may be as recent as AD 1550 and AD 1910, respectively. The sites in Frontier and Harlan cos. document the presence of white-tailed jackrabbits in the southwestern and south-central, respectively, parts of Nebraska during the late Holocene. None of these sites is more than 2000 years old (25FT18) and the Harlan Co. site could be as late as AD 1750. These data do support Jones' supposition about the southern distribution of white-tailed jackrabbits in Nebraska prior to European settlement.

Lepus sp.—jackrabbits

Site records (28).—Boone Co.: Beaver Creek Site, 25BO23, 4.5 mi. S Petersburg (Koch and Nelson, 2002: 86). Buffalo Co.: 25BF109, Sec. 4, T9N, R16W (Ludwickson, 1978b: 34); 25BF140, Sec. 35, T11N, R13W (Ludwickson, 1978b: 34). Burt Co.: Logan Creek Site A, C, 25BT3, Logan Creek, SW Oakland (Snyder and Bozell, 1983: 16; Graham et al., 1987: 283; Widga, 2006: 67). Franklin Co.: Lost Creek Site, 25FR6, Lost Creek, 2.25 mi. S, 1.5 mi. E Bloomington (Strong, 1935: 100). Frontier Co.: 25FT22, Sec. 11, T5N, R26W (Turnmire, 1996: 189-191); Mowry Bluff Site, 25FT35, 6 mi. NNW Cambridge (Falk, 1969a: 42); Lime Creek Site, 25FT41, north side of Lime Creek valley, 1 mi. from original junction of Lime and Medicine creeks (Davis, 1962: 23; Wedel, 1986: 69; Graham et al., 1987: 283); Allen Site, 25FT50, ≈ 3 mi. upstream from Medicine Creek Dam (Bamforth, 2002: 65; Hudson, 2007: 195). Garden Co.: Ash Hollow Cave Site, 25GD2, 3 mi. SE Lewellen (Champe, 1946: 43-44). Gosper Co.: Wallace Site, 25GO2, Plum Creek, 5.5 mi. N, 9.5 mi. E Elwood (Winfrey, 1991: 82). Hall Co.: Hulme Site, 25HL28, 16 mi. W Grand Island (Bozell, 1991: 233). Harlan Co.: Graham Ossuary, 25HN5, 4.25 S, 4.5 E Alma; (Adair and Brown, 1987: 569); 25HN36, 2 mi. S Alma (Adair and Brown, 1987: 153, 582). Hooker Co.: Humphrey Site, 25HO21, Middle Loup River, about 5 mi. E Mullen (Gunnerson, 1960: 204); Kelso Site, 25HO23, Middle Loup River, about 5 mi. E Mullen (Kivett, 1952: 39-40). Loup Co.: Royal Goodenow Site, 25LP8, confluence of Dry Creek and Calamus River, 13 km NW Burwell (Roper, 1989: 313-316). Morrill Co.: Greenwood Site, old Greenwood Stage Station, Keenan Ranch, ≈ 9 mi. S, 8 mi. E Redington (Renaud, 1933: 14). Nance Co.: 25NC8, 10, vicinity of Fullerton (Mick, 1983: 159, 162; Graham et al., 1987: 284); Palmer

Locality [in part], 25NC29, 5.5 mi. S, 19 mi. W Fullerton (Meadow and Peterson, 2001: 156). Sherman Co.: Bill Packer Site, 25SM9, along Davis Creek in extreme northeastern corner of county (Graham et al., 1987: 284; Bozell and Rogers, 1989: 27). Sioux Co.: 25SX157, Agate Fossil Beds National Monument, 19.25 mi. S, 5.75 mi. E Harrison (Bozell, 1993: 61; Bozell, 1994a: 47-48, 62); 25SX163, Agate Fossil Beds National Monument, 19.25 mi. S, 7.25 mi. E Harrison (Bozell, 1993: 63); 25SX268, Agate Fossil Beds National Monument (Galindo, 1997: 122). Valley Co.: Mira Creek Site, 25VY1, 6 mi. W, 3/4 mi. N North Loup (Hill and Kivett, 1940: 183). Washington Co.: Fort Atkinson Site, 25WN9, 0.5 mi. E Fort Calhoun (Carlson, 1979: 124).

Remarks.—The jackrabbits in Nebraska are identified based on external and cranial characteristics, with white-tailed jackrabbit being on average larger than the black-tailed jackrabbit (Jones, 1964; Genoways et al., 2008b). However, there is enough individual, secondary sexual, and age variation to make it impossible to separate jackrabbits based on broken cranial material and on post-cranial skeletons, which has resulted in some authors trying to use DNA to identify their rabbit subfossils (Yang et al., 2005).

ORDER SORICOMORPHA

Blarina brevicauda—northern short-tailed shrew

Site records (3).—Frontier Co.: Lime Creek Site, 25FT41, north side of Lime Creek valley, 1 mi. from original junction of Lime and Medicine creeks (Graham et al., 1987: 283). Howard Co.: Schmidt Site, 25HW301, along North Loup River, near Elba (Satorius-Fox, 1982: 8; Mick, 1983: 155; Graham et al., 1987: 284). Sarpy Co.: Lucien Fontenelle's Post, 25SY26, Bellevue (Bozell et al., 1990: 29).

Remarks.—Until 1972, all short-tailed shrews in Nebraska were assigned to this species (Jones, 1964), which was originally described from Engineer Cantonment, Washington Co., by Thomas Say (Genoways and Ratcliffe, 2008). In 1972, Genoways and Choate (1972) demonstrated that populations of short-tailed shrews in southern Nebraska represented a second species in the state. Later Benedict (1999a, 1999b; Genoways et al., 2008b) demonstrated that the two species meet along a very narrow hybrid zone extending primarily south of the Platte River from Lincoln Co. to Cass Co. This zone of hybridization is believed to have fluctuated to the north and south during the changing climatic conditions throughout the Pleistocene and Holocene (Jones et al., 1984). The northern short-tailed shrew occurs to the north of the zone of hybridization and east of the Sand Hills.

Fortunately, all records of the northern short-tailed shrew from the Holocene of Nebraska were reported after the recognition of two species in the state in 1972. Holocene records from Howard and Sarpy cos. fall within the modern distributional range of *B. brevicauda*. However, the material from the Lime Creek site in Frontier Co., is from about 24 km [15 mi.] south of where the zone of hybridization is believed to occur (Benedict, 1999b). This material is from at least 8000 years ago. This helps document that the geographic range of this species has been shifting through much of the Holocene.

Blarina hylophaga—Elliot's short-tailed shrew

Site record (1).—Hall Co.: Hulme Site, 25HL28, 16 mi. W Grand Island (Bozell, 1991: 233; Souders, 1994: 53).

Remarks.—The name *B. carolinensis* was first applied to this species (Genoways and Choate 1972), but subsequent work demonstrated that the name *B. hylophaga* was the appropriate name for this taxon. This is a species primarily of the grasslands of the plains from Nebraska and Iowa southward into Texas (George et al, 1981; Genoways et al., 2008b; Thompson et al., 2011).

The one Holocene record for Elliot's short-tailed shrew from Nebraska is from about 40 km [25 mi.] north of the modern distributional limit of this species. Specimens from the Hulme Site are at least 800 years old. This record and that of *B. brevicauda* provide further evidence of the shifting position of these populations during the Holocene.

Cryptotis parva—least shrew

Site record (1).—Burt Co.: Logan Creek Site B, 25BT3, Logan Creek, SW Oakland (Snyder and Bozell, 1983: 16: Graham et al., 1987: 283; Widga, 2006: 67).

Remarks.—This Holocene record from about 6500 years ago falls well within the modern geographic range of the least shrew (Genoways et al., 2008b; Jones, 1964).

Scalopus aquaticus—eastern mole

Site records (11).—Cass Co.: Walker Gilmore, 25CC28, 6 mi. SE Murray (Haas, 1983: 251; Graham et al., 1987: 284). Douglas Co.: Cabanne's Post, 25DO8, near Dodge Park northeast of Omaha (Bozell et al., 1990: 32). Frontier Co.: 25FT18, near point where Lime Creek originally entered Medicine Creek (Kivett and Metcalf, 1997: 212); Allen Site, 25FT50, ≈ 3 mi. upstream from Medicine Creek Dam (Hudson, 2007: 195). Gosper Co.: Wallace Site, 25GO2, Plum Creek, 5.5 mi. N, 9.5 mi. E Elwood (Winfrey, 1991: 82; Souders, 1994: 53). Howard Co.: Schmidt Site, 25HW301, along North Loup River, near Elba (Satorius-Fox, 1982: 8; Mick, 1983: 155; Graham et al., 1987: 284). Loup Co.: Royal Goodenow Site, 25LP8, confluence of Dry Creek and Calamus River, 8 mi. NW Burwell (Falk and Pepperl, 1981: A1-81); Eldon Larson Site, 25LP19, ≈ 6.5 mi. upstream from Calamus Dam; age—AD 1650-1750 (Falk and Pepperl, 1981: A1 173). Sarpy Co.: Patterson Site, 25SY31, ≈ 7 mi. S Gretna (Bozell and Ludwickson, 1999: 83). Sioux Co.: 25SX157, Agate Fossil Beds National Monument, 19.25 mi. S, 5.75 mi. E Harrison (Bozell, 1994a: 47, 62); 25SX476, Agate Fossil Beds National Monument, 1.7 mi. E Agate (Bozell, 1993: 66).

Remarks.—The eastern mole is statewide in distribution, with the possible exception of a small area of uplands in extreme western Nebraska between the North Platte and Niobrara rivers (Jones, 1964; Genoways et al., 2008b). However, Nebraska is near the western distributional limit for the species because there are only a few records from eastern Wyoming and southwestern South Dakota (Hall, 1981).

Because the eastern mole is statewide in distribution, all Holocene records fall within the modern distribution of the species. However, the records from the Agate Fossil Beds National Monument, Sioux Co., are of interest because they are from the valley of the Niobrara River just at the northern edge of the uplands where Jones (1964) reports the species to be absent. These records also are very near the western edge of the geographic range of the species and indicate that this species has been in this area for at least 3000 years (25SX157). The oldest Holocene site where eastern moles have been found in Nebraska is the Allen Site in Frontier Co., where the fauna is believed to date from at least 9500 years ago.

ORDER CARNIVORA

Puma concolor—mountain lion

Site record (1).—Frontier Co.: 25FT17, vicinity of Medicine Creek Dam (Mick, 1983: 191; Graham et al., 1987: 286; Kivett and Metcalf, 1997: 213).

Remarks.—Mountains lions originally occurred throughout Nebraska, but by the early 1900s the species had been extirpated from the state (Jones, 1964; Genoways et al., 2008b). Beginning in the 1990s wandering individuals began to re-invade, primarily in the Pine Ridge area at first, but now they can be expected anywhere in the state (Benedict et al., 2000; Hoffman and Genoways, 2005).

It is surprising to have but a single Holocene record for the mountain lion in Nebraska. It is a large and widespread species that should preserve well in paleontological and archeological sites. The single site is in Frontier Co. in the southwestern part of the state and dates from about 800 years ago.

Lynx rufus—bobcat

Site records (8).—Butler Co.: Linwood Site, 25BU1, Skull Creek, 1 mi. SW Linwood (Carlson, 1973: 79). Frontier Co.: 25FT17, vicinity of Medicine Creek Dam (Mick, 1983: 191; Graham et al., 1987: 286); 25FT77, Red Willow Creek, ≈1 mi. upstream from Red Willow Dam (Grange, 1980: 171). Morrill Co.: Greenwood Site, old Greenwood Stage Station, Keenan Ranch, ≈ 9 mi. S, 8 mi. E Redington (Renaud, 1933: 14). Nance Co.: 25NC10, vicinity of Fullerton (Mick, 1983: 162; Graham et al., 1987:285). Sherman Co.: Bill Packer Site, 25SM9, along Davis Creek in extreme northeastern corner of county (Graham et al., 1987: 285; Bozell and Rogers, 1989: 28). Sarpy Co.: Lucien Fontenelle's Post, 25SY26, Bellevue (Bozell et al., 1990: 30). Webster Co.: Hill Site, 25WT7, 2 mi. S, 7 mi. E Red Cloud (Wedel, 1936: 62).

Remarks.—Modern populations of bobcats are relatively common in Nebraska, but because of their secretive habits, individuals are seldom observed. The current distribution of the species is statewide, with larger populations in more remote areas of western Nebraska, along the Niobrara River, and other areas with dense cover (Jones, 1964; Benedict et al., 2000; Genoways et al., 2008b). These animals are predators on a wide range of mammals and birds.

Even though there are only eight Holocene sites where bobcat remains have been recovered, these sites are scattered throughout Nebraska supporting the supposition that the species was statewide in distribution during this time. However, the sites are all of relatively recent origin with the oldest site (25FT77) being from about 1500 years ago.

Lynx sp.—bobcat/lynx

Site record (1).—Frontier Co.: 25FT17, vicinity of Medicine Creek Dam (Kivett and Metcalf, 1997: 213).

Remarks.—Material from one site in southwestern Nebraska could not be assigned to a specific species. In addition to the bobcat, there are scattered records of the Canadian lynx, *Lynx canadensis*, recorded from Nebraska so there is potential that this record may apply to either species (Hoffman and Genoways, 2005).

Canis familiaris—dog

Site records (31).—Antelope Co.: 25AP37, northwestern edge of Neligh (Buhta et al., 2007: 138). Butler Co.: Linwood Site, 25BU1, Skull Creek, 1 mi. SW Linwood (Carlson, 1973: 79); Barcal Site, 25BU4, near Abie (Bozell, 1988: 98). Cass Co.: Walker Gilmore, 25CC28, 6 mi. SE Murray (Haas, 1983: 251; Graham et al., 1987: 284). Colfax Co.: Schuyler Site [Gray-Wolfe Site], 25CX1, along Shell Creek, ≈ 3 mi. NW Schuyler (Dunlevy, 1936: 199). Dakota Co.: Big Village Site, 25DK5, 1 mi. NE Homer (O'Shea and Ludwickson, 1992: 331). Franklin Co.: Lost Creek Site, 25FR6, Lost Creek, 2.25 mi. S, 1.5 mi. E Bloomington (Strong, 1935: 100). Frontier Co.: 25FT13, 16, 17, vicinity of Medicine Creek Dam (Mick, 1983: 184; Graham et al., 1987: 284; Kivett and Metcalf, 1997: 213); 25FT30, near point where Lime Creek originally entered Medicine Creek (Kivett and Metcalf, 1997: 213); Mowry Bluff Site, 25FT35, 6 mi. NW Cambridge (Falk, 1969b: 48; Mick, 1983: 205; Graham et al., 1987: 286); 25FT36, ≈ 4 km upstream on Medicine Creek from Medicine Creek Dam (Kivett and Metcalf, 1997: 213); 25FT39, ≈ 5 km upstream on Medicine Creek from Medicine Creek Dam (Kivett and Metcalf, 1997: 213). Gosper Co.: Stillman Site, 25GO35, about 15 mi. S, 8 mi. W Elwood (Philips, 1960b: 2); Dawson Site, 25GO88, Sec. 12, T5N, R22W (Philips, 1960d: 3). Hall Co.: Hulme Site, 25HL28, 16 mi. W Grand Island (Bozell, 1991: 234). Holt Co.: Redbird Site, 25HT3, 0.2 mi. N Redbird (Wood, 1956: 45). Hooker Co.: Humphrey Site, 25HO21, Middle Loup River, about 5 mi. E Mullen (Gunnerson, 1960: 204); Kelso Site, 25HO23, Middle Loup River, about 5 mi. E Mullen (Kivett, 1952: 39-40). Howard Co.: Schmidt Site, 25HW301, along North Loup River, near Elba (Mick, 1983: 156; Graham et al., 1987: 284). Knox Co.: Ponca Fort, 25KX1, 1 mi. E Verdel (Wood, 1993: 40, 64); Minarik Site, 25KX2, Ponca Creek, 7 mi. NW Verdel (Wood, 1956: 138). Nance Co.: Burkett Site, 25NC1, 4 mi. SW Genoa (Blackman, 1907: 339, 341; Dunlevy, 1936: 200); Horse Creek Site, 25NC2, near Fullerton (Bozell, 1988: 98; Hill, 1927); Wright Site, 25NC3, near Genoa (Bozell, 1988: 98; Wedel, 1938). Polk Co.: Clarks Site, 25PK1, Sec. 17, T14N, R4W (Blackman, 1907: 358; Bozell, 1988: 98). Richardson Co.: Leary Village Site, 25RH1, 4 mi. SE Rulo (Hill and Wedel, 1936: 63). Stanton Co.: Stanton Site, 25ST1, 1 mi. E Stanton (Gunnerson, n.d.). Valley Co.: Mira Creek Site, 25VY1, 6 mi. W, 3/4 mi. N North Loup (Hill and Kivett, 1940: 183). Webster Co.: Hill Site, 25WT1, 2 mi. S, 7 mi. E Red Cloud (Wedel, 1936: 62).

Remarks.—The domestic dog entered North America in company with the forebearers of Native Americans (Morey, 1994). Intentional burials and other material that have been studied indicate that the domestic dog made this journey more than 9500 years ago (Ezzo and Stiner, 2000; Morey and Wiant, 1992; Olsen, 1974; Than, 2011). The Holocene records for the domestic dog in Nebraska are concentrated in the eastern half of the state, with two sites in Hooker Co. being the western-most.

None of the dog material is from early Holocene sites, with the oldest material coming from two sites that are no more than 2100 years old—25HO23 and 25VY1. Material from three other sites could be more than 1100 years old—25CC28, 25GO88, and 25RH1. The remaining sites are nearly equally divided between those that fall between AD 1000 to AD 1500—25FR6; 25FT13, 16, 17, 30, 35, 36, 39; 25GO35; 25HL28; and 25HW301—and the remaining 15 sites that are all post AD 1500 in age.

***Canis latrans*—coyote**

Site records (28).—Butler Co.: Linwood Site, 25BU1, Skull Creek, 1 mi. SW Linwood (Carlson, 1973: 79). Cass Co.: Walker Gilmore, 25CC28, 6 mi. SE Murray (Haas, 1983: 252). Frontier Co.: 25FT13, 16, 70 vicinity of Medicine Creek Dam (Mick, 1983: 184; Graham et al., 1987: 284; Kivett and Metcalf, 1997: 213); 25FT18, 30, near point where Lime Creek originally entered Medicine Creek (Mick, 1983: 202; Graham et al., 1987: 286; Kivett and Metcalf, 1997: 212); Lime Creek Site, 25FT41, north side of Lime Creek valley, 1 mi. from original junction of Lime and Medicine creeks (Graham et al., 1987: 283); Red Smoke Site, 25FT42. north side of Lime Creek valley, 1.3 mi. from original junction of Lime and Medicine creeks (Davis and Schultz, 1952: 289; Knudson, 2002: 84); Allen Site, 25FT50, ≈ 3 mi. upstream from Medicine Creek Dam (Holder and Wike, 1949: 261; Hudson, 2007: 195); 25FT54, Red Willow Creek, 5 mi. upstream from Red Willow Dam (Grange, 1980: 171; Mick, 1983: 168); 25FT58, 80, Red Willow Creek, ≈4.5 mi. upstream from Red Willow Dam (Grange, 1980: 171; Mick, 1983: 168). Garden Co.: O. V. Clary site, 25GD50, Ash Hollow Draw, 4.25 mi. S, 1 mi. W Lewellen (Hill et. al., 2008: 124). Gosper

Co.: Wallace Site, 25GO2, Plum Creek, 5.5 mi. N, 9.5 mi. E Elwood (Winfrey, 1991: 82). Harlan Co.: 25HN36, 2 mi. S Alma (Adair and Brown, 1987: 153, 583). Howard Co.: Schmidt Site, 25HW301, along North Loup River, near Elba (Mick, 1983: 156; Graham et al., 1987: 284). Morrill Co.: Greenwood Site, old Greenwood Stage Station, Keenan Ranch, ≈ 9 mi. S, 8 mi. E Redington (Renaud, 1933: 14). Nance Co.: Burkett Site, 25NC1, 4 mi. SW Genoa (Dunlevy, 1936: 200); 25NC8, 10, vicinity of Fullerton (Mick, 1983: 159, 162; Graham et al., 1987: 284). Red Willow Co.: 25RW22, 3 mi. N, 6.25 mi. W Indianola (Grange 1980: 172; Mick, 1983: 168); Doyle Site, 25RW28, 9.5 mi. N, 0.75 mi. W McCook (Grange, 1980: 173); Gillen Pits, 25RW101, 4.5 mi. W McCook (Corner, 1977: 79-80); Davidson Pits, 25RW102, 1 mi. W Bartley (Corner, 1977: 79-80). Richardson Co.: Leary Village Site, 25RH1, 4 mi. SE Rulo (Hill and Wedel, 1936: 63). Sherman Co.: Bill Packer Site, 25SM9, along Davis Creek in extreme northeastern corner of county (Graham et al., 1987: 284). Stanton Co.: Stanton Site, 25ST1, 1 mi. E Stanton (Gunnerson, n.d.).

Remarks.—The coyote currently is statewide in distribution in Nebraska and a common inhabitant of most areas of the state (Jones, 1964; Genoways et al., 2008b). Holocene records also seem to be located in most parts of the state with the possible exceptions of the Sand Hills and along the Niobrara River in the far north.

The coyote has been present in Nebraska throughout the Holocene. Its remains have been found at several of the oldest sites, including the late Pleistocene-early Holocene transition fauna from Gillen Pits (25RW101) and Davidson Pits (25RW102) in Red Willow Co. Three early sites in Frontier Co. all included coyote remains—Allen site (25FT50), as much as 13,000 years old; Red Smoke site (25FT42), 10,000 years old; Lime Creek site (25FT41), 9500 years old. At the 9000-year-old O. V. Clary site in Garden Co., coyote remains were found in association with a Paleoindian hearth (Hill et al., 2008).

Canis lupus nubilus—gray wolf

Site records (16).—Burt Co.: Logan Creek Site B, D, 25BT3, Logan Creek, SW Oakland (Snyder and Bozell, 1983: 17-19). Franklin Co.: Lost Creek Site, 25FR6, Lost Creek, 2.25 mi. S, 1.5 mi. E Bloomington (Strong, 1935: 96). Frontier Co.: 25FT17, vicinity of Medicine Creek Dam (Mick, 1983: 190; Graham et al., 1987: 286; Kivett and Metcalf, 1997: 213); Allen Site, 25FT50, ≈ 3 mi. upstream from Medicine Creek Dam (Hudson, 2007: 195). Hall Co.: Hulme Site, 25HL28, 16 mi. W Grand Island (Bozell, 1991: 234). Howard Co.: Schmidt Site, 25HW301, along North Loup River, near Elba (Mick, 1983: 156; Graham et al., 1987: 284). Lancaster Co.: Schrader Site, 25LC1, Salt Creek, near Roca, about 9 mi. S Lincoln (Hill and Cooper, 1936a: 246). Morrill Co.: Greenwood Site, old Greenwood Stage Station, Keenan Ranch, ≈ 9 mi. S, 8 mi. E Redington (Renaud, 1933: 14); Site 21, 2.5 mi. SW Mud Springs (Bell and Cape, 1936: 383); Site 45, ≈ 9 mi. S, 8 mi. E Redington (Bell and Cape, 1936: 382). Nance Co.: Burkett Site, 25NC1, 4 mi. SW Genoa (Dunlevy, 1936: 200). Red Willow Co.: Gillen Pits, 25RW101, 4.5 mi. W McCook (Corner, 1977: 79-80); Davidson Pits, 25RW102, 1 mi. W Bartley (Corner, 1977: 79-80). Richardson Co.: Leary Village Site, 25RH1, 4 mi. SE Rulo (Hill and Wedel, 1936: 63). Stanton Co.: Stanton Site, 25ST1, 1 mi. E Stanton (Gunnerson, n.d.).

Remarks.—This subspecies of the gray wolf was the hunter of bison on the Great Plains. It was originally described from Engineer Cantonment, Washington Co. (Genoways and Ratcliffe, 2008) and was statewide in distribution. Many early authors comment that the wolves were abundant in the state (Jones, 1964; Genoways et al., 2008b). The gray wolf was gone from Nebraska by 1920 and this subspecies is now considered to be extinct. Surprisingly, the taxonomy of wolves is currently in a state of flux, but all definitions of populations still include *C. l. nubilus*. Some authors give the buffalo wolf a geographic range far beyond the Great Plains, while others only recognize their presence on the Great Plains. In the latter scenario, this wolf is considered to be extinct, whereas in the former some living populations are considered to exist. A recent record of a wolf from Greeley Co. was assigned to *C. l. lycaon*, which is a subspecies that lies to the north in Minnesota and surrounding areas (Hoffman and Genoways, 2005).

Although there are only 16 Holocene records of the gray wolf in Nebraska, the records are scattered to many areas of the state. Morrill Co. is in the Panhandle, whereas Richardson Co. is the southeastern-most county and Lancaster Co. also would be considered southeastern. In the southwestern corner there are records from Frontier and Red Willow cos. and Franklin Co. is in south-central Nebraska. Burt Co. is along the eastern border and to the west in east-central Nebraska are several counties with records—Hall, Howard, Nance and Stanton cos. Areas lacking records include northeastern and northern Nebraska and the Sand Hills.

Gray wolves are known from two of the Pleistocene/Holocene transition localities in Red Willow Co. in southwestern Nebraska so the species was present throughout the Holocene along with their primary prey species, the bison. Also in southwestern Nebraska is the early Holocene Allen site in Frontier Co. with material that may be 13,000 years old. In far eastern Nebraska, remains of the wolves were discovered at the Logan Creek site in Burt Co., where deposits are aged to 6000 to 7000 years old. The remaining sites are less than 1300 years old.

Canis sp.—dogs and relatives

Site records (52).—Boone Co.: Beaver Creek Site, 25BO23, 2.4 km N Loretto (Koch, 2002: i, 1-4; Koch and Nelson, 2002: 87). Brown Co.: McIntosh Site, 25BW15, Enders Lake, ≈12 mi. S, 12.5 mi. W Ainsworth (Koch, 1995: 43; 2004: 117). Buffalo Co.: Post Draw Site, 25BF101, Sec. 35, T11N, R13W (Ludwickson, 1978b: 34). Burt Co.: Logan Creek Site C, D, 25BT3, Logan Creek, SW Oakland (Graham et al., 1987: 283; Widga, 2006: 67-68). Butler Co.: Barcal Site, 25BU4, near Abie (O'Shea and Ludwickson, 1992: 338). Cass Co.: Walker Gilmore, 25CC28, 6 mi. SE Murray (Haas, 1983: 251; Graham et al., 1987: 284); Little Pawnee Creek Site, 25CC228, 0.8 mi. N, 3 mi. W South Bend (Ludwickson, 1994:116). Cedar Co.: Schulte Site, 25CD1, St. Helena (Cooper, 1936: 76); Wiseman Site, 25CD3, 4.5 mi. NE Wynot (Cooper, 1936: 76); Ferber Site, 25CD10, near Bow Valley Mill (Ludwickson et al., 1981: 22-23). Colfax Co.: Schuyler Site, 25CX1, along Shell Creek, ≈ 3 mi. NW Schuyler (Dunley, 1936: 199; Bozell et al., 1982: 27; Graham et al., 1987: 286). Chase Co.: Lovitt Site, 25CH1, 12 mi. W Wauneta (Hill and Metcalf, 1942: 204; Gunnerson, 1960: 212-216). Dakota Co.: Bobier Site, 25DK1, 4.5 mi. SE Homer (Frantz, 1963: 88-89); Big Village Site, 25DK5, 1 mi. NE Homer (O'Shea and Ludwickson, 1992: 331). Dixon Co.: Annie's Site, 25DX30, in vicinity of 2 mi. N, 3 mi. E Newcastle (Manz and Blakeslee, 1988: 159-160). Douglas Co.: Parker Site, 25DO2, on the bluffs north of Ponca Creek and west of the Missouri River, in extreme northeastern Douglas Co. (Ewing, 2000: 4); Cabanne's Post, 25DO8, near Dodge Park northeast of Omaha (Bozell et al., 1990: 32). Frontier Co.: 25FT13, 17, 70, vicinity of Medicine Creek Dam (Mick, 1983: 174, 190, 210; Graham et al., 1987: 284, 286); 25FT22, Sec. 11, T5N, R26W (Turnmire, 1996: 189-191); 25FT30, near point where Lime Creek originally entered Medicine Creek (Mick, 1983: 201; Graham et al., 1987: 286); Spring Creek Site, 25FT31, Red Willow Creek, adjacent to Red Willow Dam (Grange, 1980: 166); 25FT36, ≈ 4 km upstream on Medicine Creek from Medicine Creek Dam (Mick, 1983: 206); 25FT39, ≈ 5 km upstream on Medicine Creek from Medicine Creek Dam (Mick, 1983: 210; Graham et al., 1987: 286); Lime Creek Site, 25FT41, north side of Lime Creek valley, 1 mi. from original junction of Lime and Medicine creeks (Davis, 1962: 23; Wedel, 1986: 69). Garfield Co.: 25GF13, 1 mi. S, 2.1 mi. W Burwell (Bozell and Pepperl, 1982: 10, 12). Garden Co.: Ash Hollow Cave Site, 25GD2, 3 mi. SE Lewellen (Champe, 1946: 43). Gosper Co.: Wallace Site, 25GO2, Plum Creek, 5.5 mi. N, 9.5 mi. E Elwood (Winfrey, 1991: 82); Golter Site, 25GO30, 4 mi. N Arapahoe [Furnas Co.] (Philips, 1957b: 1). Hall Co.: Hulme Site, 25HL28, 16 mi. W Grand Island (Bozell, 1991: 234). Harlan Co.: White Cat Village, 25HN37, Prairie Dog Creek, 6 mi. SE Alma (Champe, 1949: 289; Gunnerson, 1960: 177; Adair and Brown, 1987: 168, 588). Hooker Co.: Kelso Site, 25HO23, Middle Loup River, about 5 mi. E Mullen (Kivett, 1952: 39-40). Howard Co.: Stabaco Site, 25HW16, Loup River, near Cushing (Goedert, 1995: 170); Schmidt Site, 25HW301, along North Loup River, near Elba (Mick, 1983: 156; Graham et al., 1987: 284). Keya Paha Co.: 25KP243, East Middle Creek, 2.25 mi. S, 1.25 mi. E Norden (Pepperl and Falk, 1983: B91). Loup Co.: Royal Goodenow Site, 25LP8, confluence of Dry Creek and Calamus River, 8 mi. NW Burwell (Falk and Pepperl, 1981: A1-81; Roper, 1989: 313-316). Merrick Co.: Tahaksu Site, 25MK15, 4.8 mi. N, 1.2 mi. W Palmer (Watson, 1996: 135). Nance Co.: Palmer Locality [in part], 25NC29, 5.5 mi. S, 19 mi. W Fullerton (Meadow and Peterson, 2001: 157). Nemaha Co.: 25OT24, 2.5 mi. S, 0.75 mi. E Brownsville (Kaldahl, 1993: 109); Majors Site, 25NH2, ≈ 3 mi. NW Peru (Hill and Cooper, 1937: 324).

Red Willow Co.: 25RW22, 3 mi. N, 6.25 mi. W Indianola (Grange 1980: 166; Mick, 1983: 168). Richardson Co.: Rulo Southeast, 25RH69, 0.4 km S Rulo (Nelson, 2006: 165). Sarpy Co.: Lucien Fontenelle's Post, 25SY26, Bellevue (Bozell et al., 1990: 29); Patterson Site, 25SY31, ≈ 7 mi. S Gretna (Bozell and Ludwickson, 1999: 83). Scotts Bluff Co.: 25SF76, Scotts Bluff National Monument, 0.7 mi. N, 0.1 mi. W Visitor Center (Bozell, 1994b: 52, 63). Sherman Co.: Bill Packer Site, 25SM9, along Davis Creek in extreme northeastern corner of county (Graham et al., 1987: 284; Bozell and Rogers, 1989: 28). Sioux Co.: 25SX163, Agate Fossil Beds National Monument, 19.25 mi. S, 7.25 mi. E Harrison (Bozell, 1993: 63-64); 25SX475, Agate Fossil Beds National Monument, 1.5 mi. E Agate (Bozell, 1993: 66). Washington Co.: Fort Atkinson Site, 25WN9, 0.5 mi. E Fort Calhoun (Carlson, 1979: 124). Webster Co.: Hill Site, 25WT1, 2 mi. S, 7 mi. E Red Cloud (Metcalf, 1941b: 30); Shipman Site, 25WT7, south side of Republican River, between Red Cloud and Guide Rock (Mick, 1983: 154; Graham et al., 1987: 286).

Remarks.—With three members of the genus *Canis* possibly present at all sites within Nebraska, post-cranial and fragmentary materials are impossible to identify to the species-level using current methods.

Urocyon cinereoargenteus—gray fox

Site records (5).—Chase Co.: Lovitt Site, 25CH1, 12 mi. W Wauneta (Hill and Metcalf, 1942: 204). Frontier Co.: 25FT16, 17, 70, vicinity of Medicine Creek Dam (Mick, 1983: 184, 190, 210; Graham et al., 1987:285-286; Kivett and Metcalf, 1997: 213); 25FT18, near point where Lime Creek originally entered Medicine Creek (Kivett and Metcalf, 1997: 212).

Remarks.—Modern populations of the gray fox are known in eastern Nebraska and westward in the Platte River drainage (Jones, 1964; Genoways et al., 2008b) and along the Niobrara River in northern Nebraska (Benedict et al., 2000); however, there are no records from the southwestern corner of the state. On the other hand, all five of the Holocene sites from which the gray fox has been reported are from two counties in the southwestern corner of the state. The oldest remains come from 25FT18 in Frontier Co. and are about 1500 years old, whereas the most recent remains are from the Lovitt site in Chase Co. and are about 300 years old. These data would seem to indicate that the gray fox occurred in southwestern Nebraska for well over 1000 years, but sometime in the last 300 years they disappeared from the area. This disappearance may have been related to habitat changes associated with European settlement of the area and the introduction of agriculture.

Vulpes velox—swift fox

Site records (14).—Boone Co.: Beaver Creek Site, 25BO23, 2.4 km N Loretto (Koch, 2002: i, 1-4; Koch and Nelson, 2002: 87). Brown Co.: McIntosh Site, 25BW15, Enders Lake, ≈12 mi. S, 12.5 mi. W Ainsworth (Koch, 1995: 43; 2004: 117). Frontier Co.: 25FT13, 16, vicinity of Medicine Creek Dam (Kivett and Metcalf, 1997: 213); 25FT30, near point where Lime Creek originally entered Medicine Creek (Mick, 1983: 202; Graham et al., 1987: 286; Kivett and Metcalf, 1997: 213); Spring Creek Site, 25FT31, Red Willow Creek, adjacent to Red Willow Dam (Grange, 1980: 166; Widga, 2004: 29); Mowry Bluff Site, 25FT35, 6 mi. NW Cambridge (Falk, 1969b: 48; Mick, 1983: 205; Graham et al., 1987: 286); Allen Site, 25FT50, ≈ 3 mi. upstream from Medicine Creek Dam (Hudson, 2007: 195); 25FT54, Red Willow Creek, 5 mi. upstream from Red Willow Dam (Grange, 1980: 171; Mick, 1983: 168). Gosper Co.: Wallace Site, 25GO2, Plum Creek, 5.5 mi. N, 9.5 mi. E Elwood (Winfrey, 1991: 82). Hall Co.: Hulme Site, 25HL28, 16 mi. W Grand Island (Bozell, 1991: 234). Harlan Co.: 25HN36, 2 mi. S Alma (Adair and Brown, 1987: 153, 584). Howard Co.: Schmidt Site, 25HW301, along North Loup River, near Elba (Mick, 1983: 156; Graham et al., 1987: 284). Morrill Co.: Greenwood Site, old Greenwood Stage Station, Keenan Ranch, ≈ 9 mi. S, 8 mi. E Redington (Renaud, 1933: 14).

Remarks.—The swift fox is the species of fox best represented in the archeological record in Nebraska, being recorded at 14 sites. Jones (1964; Genoways et al., 2008b) believed that the swift fox may have been extirpated in Nebraska by the early twentieth century, with the species re-entering the western-most parts of the state in the 1950s. Jones (1964) had swift fox historic geographic range occurring as far east as a line connecting Antelope, Madison, and Saline cos., based on historic records.

The Holocene records support such an eastern limit with records from just west (within the historic range) of the line in Boone, Brown, Hall, and Howard cos. The age of the remains from these sites all fall in the range between 450 to 900 years ago. There are records from three sites in the southwestern counties of Frontier, Gosper, and Harlan. Finally, there are remains of the swift fox from a site in the western Panhandle in Morrill Co. The oldest remains of swift fox in Nebraska come from the Allen site (25FT50) in Frontier Co., where the material comes from between 9500 to 13,000 years old.

***Vulpes fulva*—red fox**

Site records (2).—Colfax Co.: Schuyler Site, 25CX1, along Shell Creek, ≈ 3 mi. NW Schuyler (Bozell et al., 1982: 27; Graham et al., 1987: 286). Red Willow Co.: Davidson Pits, 25RW102, 1 mi. W Bartley (Corner, 1977: 79-80).

Remarks.—Although the red fox is the commonest species of fox in the modern fauna of Nebraska, it is represented by remains at only two sites in the Holocene record. Jones (1964; Genoways et al., 2008b) indicated that the red fox was common in eastern Nebraska, but was absent from the southwestern and northwestern corners of the state. The Pleistocene and Holocene biogeography of the red fox in North America is extremely complex and only in recent years has it become more clearly defined using genetic data (Aubry et al., 2009; Statham et al., 2012, 2014). The Davidson Pit site in Red Willow Co. is in the southwestern corner of the state and is apparently the first fossil record of the red fox from the late Pleistocene/early Holocene boundary on the Great Plains (Corner, 1977). At this time red fox populations in North America were believed to be isolated south of the ice sheets in two forest refugia. One of the refugia was in western mountains, whereas the other was in the eastern United States. The partial lower jaw from Davidson Pit is from an area that lies between these two refugia, but probably relates to the refugium in the western mountains as do other species from southwestern Nebraska from this time period (see Table 1 and Introduction). As the ice sheets retreated northward during the Holocene red fox populations followed them, occupying the boreal forest of Canada and Alaska and expanding their range in the western mountains. As European settlement of the eastern United States began, red foxes were absent from the area. The vacant areas were reoccupied during the Colonial era (AD 1650 to 1800) by a natural southward expansion of boreal populations in response to "the conversion of hardwood forests to farmlands" (Statham et al., 2012). The record from Colfax Co. in east-central Nebraska where the species is still considered to be common would fit the scenario described by Statham et al. (2012), except the timing is not in synchrony. These remains are no younger than AD 1600 so they are older than would be anticipated. The possibility is that red foxes arrived earlier on the Great Plains than the eastern deciduous forests. This potential seems further supported by the earliest historic record for red fox in Nebraska based on an individual figured by Titian Peale from 1819-1820 at Engineer Cantonment in Washington Co. (Genoways and Ratcliffe, 2008). At this time, there were only a few European trading posts along the Missouri River and no ground had been broken for farmland.

Roehrs et al. (2021) gave a detailed discussion of the reasons to use the scientific name *Vulpes fulva* rather than *Vulpes vulpes* for this species.

Ursus americanus—American black bear

Site records (3).—Frontier Co.: 25FT13, vicinity of Medicine Creek Dam (Kivett and Metcalf, 1997: 213); Allen Site, 25FT50, ≈ 3 mi. upstream from Medicine Creek Dam (Hudson, 2007: 195). Nance Co.: Burkett Site, 5 mi. SW Genoa (Blackman, 1907: 341).

Remarks.—These three records of the black bear are from southwestern Nebraska (2) and central Nebraska not far from the Platte River. The latter record from the Burkett Site in Nance Co. was based on "a few tusks, which may be those of the bear" according to Blackman (1907: 341).

Jones (1964) found the historical records of black bears in eastern Nebraska (Genoways et al., 2008b; Genoways and Ratcliffe, 2008) and westward along the Loup and Niobrara rivers. After a century or more of being extirpated from Nebraska black bears are entering the state along its western boundary. These bears are coming from the Rocky Mountains to the west (Hoffman and Genoways, 2005; Hoffman et al., 2009). Black bears may have been entering Nebraska from the east and the west during the Holocene. The localities in Frontier Co. are essentially equidistant (about 390 km [240 mi.] in each direction) between the Missouri River and the Rocky Mountains making the determination of the source of these bears impossible to know with the data at hand.

Ursus sp.—bear

Site records (4).—Butler Co.: Barcal Site, 25BU4, near Abie (O'Shea and Ludwickson, 1992: 338). Howard Co.: Stabaco Site, 25HW16, Loup River, near Cushing (Goedert, 1995: 170). Stanton Co.: Stanton Site, 25ST1, 1 mi. E Stanton (Gunnerson, n.d.). Webster Co.: Hill Site, 25WT1, 2 mi. S, 7 mi. E Red Cloud (Metcalf, 1941b: 51).

Remarks.—In addition to the black bear, the brown or grizzly bear, *Ursus arctos horribilis*, occurred in Nebraska during historic (Jones, 1964) as well as Holocene times so that some remains can't be determined to species.

Material from the Stabaco Site (25HW16) in Howard Co. was identified simply as cf. *Ursus* (Goedert, 1995: 170). The record from the Stanton Site (25ST1), Stanton Co., is based only on canines (Gunnerson, n.d.). These sites are both very young, being occupied 1725 to 1760 and 1800 to 1833, respectively.

By far the most interesting and important find was from the Hill Site (25WT1) in Webster Co. (Metcalf, 1941b: 51). According to Metcalf (1941b) the burial of an infant (Burial 8, Burial Hill 2) was notable for all of the included grave goods. "About the neck was a collar of bear-claws, those which were under the collar [another collar made of silver] being by far in the best state of preservation. The claws had been arranged in sets of ten, and at least three—possibly four—sets had formed the collar. Each claw except for a short one on each side of each set had been perforated in two places and between the claws at each perforation was a large blue bead, 16 to each set. No other bones of the paw were present, but at the articular end of each set of claws was a great deal of hair and an occasional scrap of hide or leather. Two sets of claws were in exceedingly poor condition, some being to a cornmeal-like condition by decay." This is also a very recent site, with occupation occurring between 1700 and 1850.

Lontra canadensis—North American river otter

Site records (4).—Burt Co.: Logan Creek Site C, 25BT3, Logan Creek, SW Oakland (Snyder and Bozell, 1983: 17; Graham et al., 1987: 283; Widga, 2006: 67). Dakota Co.: Big Village Site, 25DK5, 1 mi. NE Homer (O'Shea and Ludwickson, 1992: 331). Howard Co.: Schmidt Site, 25HW301, along North Loup River, near Elba (Mick, 1983: 157; Graham et al., 1987: 284). Sarpy Co.: Lucien Fontenelle's Post, 25SY26, Bellevue (Bozell et al., 1990: 30).

Remarks.—The North American river otter was long known under the scientific name *Lutra canadensis*, but with the work of Van Zyll de Jong (1972, 1987, 1991) and Koepfli and Wayne (1998), the use of generic name *Lutra* has been restricted to the Old World otters and the generic name *Lontra* applied to the New World otters.

Before European settlement, the North American river otter probably occurred statewide in association with aquatic habitats: however, by the early 20th century, otters had been extirpated from Nebraska (Jones, 1964; Genoways et al., 2008b). In 1986, the Nebraska Game and Parks Commission began an otter restoration program, which has successfully repopulated many of Nebraska's river systems (Hoffman and Genoways, 2005).

Holocene otter remains are known from only four sites in eastern Nebraska. All sites are associated with major aquatic habitats in Nebraska, with three from counties along the Missouri River and the fourth from a site along the North Loup River. It is surprising at how few otter remains have been found considering the uses and desirability of the pelage of this semi-aquatic mammal. The remains from Logan Creek site (25BT3) are 7000 years old, whereas those from the Schmidt site (25HW301) date from between 550 to 900 years ago. The records from the other two sites are from the last 250 years.

Mustela frenata—long-tailed weasel

Site records (5).—Frontier Co.: Owens Site, 25FT3, on Medicine Creek (Mick, 1983: 172; Graham et al., 1987: 285). Howard Co.: Schmidt Site, 25HW301, along North Loup River, near Elba (Mick, 1983: 156; Graham et al., 1987: 284). Sarpy Co.: Patterson Site, 25SY31, ≈ 7 mi. S Gretna (Bozell and Ludwickson, 1999: 83). Sherman Co.: Bill Packer Site, 25SM9, along Davis Creek in extreme northeastern corner of county (Graham et al., 1987: 285; Bozell and Rogers, 1989: 28). Washington Co.: Fort Atkinson Site, 25WN9, 0.5 mi. E Fort Calhoun (Carlson, 1979: 124).

Remarks.—The long-tailed weasel occurs statewide in Nebraska and is relatively common in some areas such as the Sand Hills (Jones, 1964; Genoways et al., 2008b). Holocene records of the long-tailed weasel are confined to southwestern, central, and eastern Nebraska, leaving most of the state without documentation of the presence of this species. With the exception of the relatively modern remains from Fort Atkinson, all other sites date from about 1000 to 1100 years ago.

Mustela nigripes—black-footed ferret

Site record (1).—Red Willow Co.: Gillen Pits, 25RW101, 4.5 mi. W McCook (Corner, 1977: 79-81).

Remarks.—The black-footed ferret is at present classified as an endangered species and is essentially extinct in the wild, although major reintroduction efforts are underway. Prior to European settlement, the black-footed ferret probably occurred throughout Nebraska with the possible exception of the eastern-most tier of counties (Jones, 1964; Genoways et al., 2008b). The species no longer occurs in the wild in Nebraska. In the past the presence and distribution of the black-footed ferret was tightly tied to its primary prey species, the black-tailed prairie dog.

Wisely et al. (2008) believed based on genetic studies that modern populations of the black-footed ferret originated from a single Pleistocene refugium on the Great Plains. There were other Pleistocene refugia for the species but the Holocene and modern populations came from this single source. The single Nebraska record from the latest Pleistocene/early Holocene indicates that the black-footed ferret was in Nebraska at the earliest point of the Holocene. However, Nebraska fossils do not record the subsequent increase in the species' distribution as the grasslands expanded in the Holocene.

***Mustela* sp.—weasel**

Site records (3).—Frontier Co.: Allen Site, 25FT50, ≈ 3 mi. upstream from Medicine Creek Dam (Hudson, 2007: 195). Douglas Co.: Parker Site, 25DO2, on the bluffs north of Ponca Creek and west of the Missouri River, in extreme northeastern Douglas Co. (Ewing, 2000: 7). Scotts Bluff Co.: 25SF53, 3.25 mi. S, 0.75 mi. E Lyman (Koch and Miller, 1996: 36).

Remarks.—In addition to the preceding two species, a third member of the genus *Mustela* currently occurs in Nebraska—the least weasel, *M. nivalis*. These species form a graded series in size with the black-footed ferret the largest and the least weasel among the smallest weasels in North America. These animals also exhibit strong secondary sexual variation with males being larger than females. These overlapping sizes make identification of skeletal elements that lack unique diagnostic characters very difficult.

The material from the Allen site (25FT50) may be as much as 13,000 years old; from 25SF53 8000 years old; and from 25DO2 between 750 and 1100 years old.

***Neovison vison*—American mink**

Site records (8).—Boone Co.: Beaver Creek Site, 25BO23, 2.4 km N Loretto (Koch, 2002: i, 1-4; Koch and Nelson, 2002: 87). Cass Co.: Walker Gilmore, 25CC28, 6 mi. SE Murray (Haas, 1983: 251; Graham et al., 1987: 285). Dixon Co.: Annie's Site, 25DX30, in vicinity of 2 mi. N, 3 mi. E Newcastle (Manz and Blakeslee, 1988: 159-160). Frontier Co.: 25FT13, vicinity of Medicine Creek Dam (Mick, 1983: 175; Graham et al., 1987: 285; Kivett and Metcalf, 1997: 213); 25FT22, Sec. 11, T5N, R26W (Turnmire, 1996: 189-191). Hall Co.: Hulme Site, 25HL28, 16 mi. W Grand Island (Bozell, 1991: 234). Hooker Co.: Humphrey Site, 25HO21, Middle Loup River, about 5 mi. E Mullen (Gunnerson, 1960: 204). Howard Co.: Schmidt Site, 25HW301, along North Loup River, near Elba (Mick, 1983: 156; Graham et al., 1987: 284).

Remarks.—The American mink long has been known under the scientific name *Mustela vison*, but the work of Abramov (2000) has presented evidence that the mink should be placed in its own unique genus, *Neovison*. American mink occur statewide in Nebraska, but are most commonly associated with aquatic habitats (Jones, 1964; Genoways et al., 2008b). The muskrat is the primary food item of the mink.

It is surprising that the American mink is known from only seven Holocene sites in Nebraska. The fur of this semi-aquatic mammal was valued for it texture and for repelling water. It was used by native people in ceremonial costumes and for various utilitarian clothing items. Obviously, their bones did not get deposited in many archeological sites.

There are no records of Holocene mink from the Panhandle of Nebraska, with the western-most sites for remains being in Hooker and Frontier cos. The other sites are scattered in central (Boone, Hall, Howard cos.), northeastern (Dixon Co.) and eastern (Cass Co.) parts of the state. These Holocene records all originated in a time period between 400 and 1300 years ago.

Gulo gulo—wolverine

Site records (3).—Frontier Co.: 25FT16, vicinity of Medicine Creek Dam (Kivett and Metcalf, 1997: 213); 25FT17, vicinity of Medicine Creek Dam (Mick, 1983: 191; Graham et al., 1987: 286; Kivett and Metcalf, 1997: 213); 25FT30, near point where Lime Creek originally entered Medicine Creek (Kivett and Metcalf, 1997: 213).

Remarks.—The wolverine is currently not part of the mammalian fauna of Nebraska, with the nearest resident population in the Rocky Mountains in northwestern Wyoming (Hoak et al., 1982; Beauvais and Johnson, 2004). The primary distribution of the species is further to the north in the boreal forests of Canada and Alaska (Hall, 1981). Jones (1964; Genoways et al., 2008b) reported an individual from near Gering, Scotts Bluff Co., taken in 1887 that is represented by a mounted specimen in the University of Nebraska State Museum and observations of this species in the Pine Ridge at about this same time and another seen near Chimney Rock, Morrill Co., in 1837. These historical records mark the known southern boundary of the wolverine distribution on the central Great Plains where the species may have been represented by wandering individuals from the Rocky Mountains to the west.

These three Holocene records, all from Frontier Co., are significant because they document a more southerly and easterly distribution of the wolverine on the Great Plains during the Holocene than previously known. The age of these sites fall in the range of 700 to 1000 years ago. Because the remains come from three sites that are geographically close to each other, it is difficult to know how many individual wolverines are involved, but I believe that it is more than one. These wolverines may have reached southwestern Nebraska from the Rocky Mountains via the Republican River and it tributaries, especially if the climate were cooler and moister than at present.

Taxidea taxus—American badger

Site records (34).—Burt Co.: Logan Creek Site B, C, 25BT3, Logan Creek, SW Oakland (Snyder and Bozell, 1983: 17; Graham et al., 1987: 283; Widga, 2006: 67). Cedar Co.: Ferber Site, 25CD10, near Bow Valley Mill (Ludwickson et al., 1981: 22-23). Colfax Co.: Schuyler Site, 25CX1, along Shell Creek, ≈ 3 mi. NW Schuyler (Bozell et al., 1982: 27; Graham et al., 1987: 286). Dawson Co.: 25DS109, 6 mi. SW Overton (Ludwickson, 1978b: 36). Douglas Co.: Cabanne's Post, 25DO8, near Dodge Park northeast of Omaha (Bozell et al., 1990: 32). Frontier Co.: 25FT13, 16, 17, 70, vicinity of Medicine Creek Dam (Mick, 1983: 175, 184, 191, 210; Graham et al., 1987: 285-286; Kivett and Metcalf, 1997: 213); 25FT39, ≈ 3.5 mi. upstream on Medicine Creek from Medicine Creek Dam (Mick, 1983: 210; Graham et al., 1987: 286; Kivett and Metcalf, 1997: 213); Lime Creek Site, 25FT41, north side of Lime Creek valley, 1 mi. from original junction of Lime and Medicine creeks (Graham et al., 1987: 283); Allen Site, 25FT50, ≈ 3 mi. upstream from Medicine Creek Dam (Hudson, 2007: 195); 25FT54, Red Willow Creek, 5 mi. upstream from Red Willow Dam (Grange, 1980: 171; Mick, 1983: 168). Gosper Co.: Wallace Site, 25GO2, Plum Creek, 5.5 mi. N, 9.5 mi. E Elwood (Winfrey, 1991: 82). Hall Co.: Hulme Site, 25HL28, 16 mi. W Grand

Island (Bozell, 1991: 234). Hooker Co.: Humphrey Site, 25HO21, Middle Loup River, about 5 mi. E Mullen (Gunnerson, 1960: 204); Kelso Site, 25HO23, Middle Loup River, about 5 mi. E Mullen (Kivett, 1952: 39-40). Howard Co.: Schmidt Site, 25HW301, along North Loup River, near Elba (Mick, 1983: 156; Graham et al., 1987: 284). Merrick Co.: Tahaksu Site, 25MK15, 4.8 mi. N, 1.2 mi. W Palmer (Watson, 1996: 135). Nance Co.: 25NC8, 10, 13, vicinity of Fullerton (Mick, 1983: 159, 162, 165; Graham et al., 1987:285); Palmer Locality [in part], 25NC29, 5.5 mi. S, 19 mi. W Fullerton (Meadow and Peterson, 2001: 157). Otoe Co.: 25OT24, Sec. 26, T8N, R14E (Kaldahl, 1993: 106-112). Red Willow Co.: Doyle Site, 25RW28, 9.5 mi. N, 0.75 mi. W McCook (Grange, 1980: 167, 173); Gillen Pits, 25RW101, 4.5 mi. W McCook (Corner, 1977: 79-80); Redwillow 1, 6 mi. NW McCook (Wedel, 1934: 171-172). Sarpy Co.: Lucien Fontenelle's Post, 25SY26, Bellevue (Bozell et al., 1990: 29); Patterson Site, 25SY31, ≈ 7 mi. S Gretna (Bozell and Ludwickson, 1999: 83). Sherman Co.: Bill Packer Site, 25SM9, along Davis Creek in extreme northeastern corner of county (Graham et al., 1987: 285; Bozell and Rogers, 1989: 28). Sioux Co.: 25SX471, Agate Fossil Beds National Monument, 0.4 mi. E Agate (Bozell, 1993: 65). Stanton Co.: Stanton Site, 25ST1, 1 mi. E Stanton (Gunnerson, n.d.). Valley Co.: Mira Creek Site, 25VY1, 6 mi. W, 0.75 mi. N North Loup (Hill and Kivett, 1940: 183). Washington Co.: Fort Atkinson Site, 25WN9, 0.5 mi. E Fort Calhoun (Carlson, 1979: 124).

Remarks.—The American badger is statewide in distribution in Nebraska, being most abundant in grasslands and uncultivated areas (Jones, 1964; Genoways et al., 2008b). There is some difficulty in interpreting these Holocene records because the badger makes large, deep burrows and some of these remains may result from intrusions into existing fossil or archeological sites.

Holocene records would support a statewide distribution for the badger during this time period. There are records from the southeastern (Otoe Co.) to northwestern (Sioux Co.) and from northeastern (Cedar Co.) to southwestern (Frontier and Red Willow cos.) parts of state. However, the primary concentration of sites is in nine counties in central to east-central parts of the state, from Dawson Co. to Stanton Co. The Sand Hills are represented by remains from the Humphrey and Kelso sites in Hooker Co.

These records also indicate that the badger was a resident of Nebraska throughout the Holocene. There are remains from some of the earliest Holocene sites in Red Willow Co., such as Gillen Pits and sites in Frontier Co. where remains may be as much as 13,000 years old—Lime Creek and Allen sites. Along the eastern boundary of Nebraska in Burt Co. (25BT3, C), the material from the Logan Creek site may be as much as 7000 years old, whereas material from central Nebraska in Valley Co. at the Mira Creek site may be as much as 2000 years old. There are also badger remains reported from the historic forts and posts—Cabanne's Post, Douglas Co.; Fort Atkinson, Washington Co.; Lucien Fontenelle's Post, Sarpy Co.

***Mephitis mephitis*—striped skunk**

Site records (20).—Banner Co.: 25BN28, Pumpkin Creek, 7 mi. NE Harrisburg (Steinacher and Carlson, 1984: 2). Boone Co.: Beaver Creek Site, 25BO23, 2.4 km N Loretto (Koch, 2002: i, 1-4; Koch and Nelson, 2002: 87). Brown Co.: McIntosh Site, 25BW15, Enders Lake, ≈12 mi. S, 12.5 mi. W Ainsworth (Koch, 1995: 43; 2004: 117). Burt Co.: Logan Creek Site B, 25BT3, Logan Creek, SW Oakland (Snyder and Bozell, 1983: 17; Graham et al., 1987: 283; Widga, 2006: 67). Colfax Co.: Schuyler Site, 25CX1, along Shell Creek, ≈ 3 mi. NW Schuyler (Bozell et al., 1982: 27; Graham et al., 1987: 286). Dakota Co.: Nelson Site, 25DK7, 1 mi. N Homer (Frantz, 1963: 104); 25DK10, 1 mi. N Homer (O'Shea and Ludwickson, 1992: 234). Douglas Co.: Cabanne's Post, 25DO8, near Dodge Park northeast of Omaha (Bozell et al., 1990: 32). Hooker Co.: Humphrey Site, 25HO21, Middle Loup River, about 5 mi. E Mullen (Gunnerson, 1960: 204). Frontier Co.: 25FT16, vicinity of Medicine Creek Dam (Mick, 1983: 185; Graham et al., 1987:285; Kivett and Metcalf, 1997: 213). Garden Co.: Ash Hollow Cave Site, 25GD2, 3 mi. SE Lewellen (Champe, 1946: 43). Gosper Co.: Wallace Site, 25GO2, Plum Creek, 5.5 mi. N, 9.5 mi. E Elwood (Winfrey, 1991: 83). Howard Co.: Schmidt Site, 25HW301, along North Loup River, near Elba (Mick, 1983: 156; Graham et al., 1987: 284). Knox Co.: Niobrara Railroad Bridge Site, 25KX207, confluence of Niobrara and

Missouri rivers (Howard and Gant, 1966: 25). Nance Co.: 25NC8, vicinity of Fullerton (Mick, 1983: 159; Graham et al., 1987:285). Richardson Co.: Leary Village Site, 25RH1, 4 mi. SE Rulo (Hill and Wedel, 1936: 63). Sarpy Co.: Lucien Fontenelle's Post, 25SY26, Bellevue (Bozell et al., 1990: 29). Sherman Co.: Bill Packer Site, 25SM9, along Davis Creek in extreme northeastern corner of county (Graham et al., 1987: 285; Bozell and Rogers, 1989: 28). Stanton Co.: Stanton Site, 25ST1, 1 mi. E Stanton (Gunnerson, n.d.). Valley Co.: Mira Creek Site, 25VY1, 6 mi. W, 0.75 mi. N North Loup (Hill and Kivett, 1940: 183).

Remarks.—The striped skunk is a common and statewide resident of Nebraska. This small carnivore consumes a wide variety of small vertebrates, insects, and other invertebrates (Jones, 1964; Genoways et al., 2008b).

The 20 Holocene records from 19 counties for the striped skunk are scattered so as to represent all regions of the state—Panhandle (Banner and Garden cos.), Sand Hills (Hooker Co.), north (Brown and Knox cos.), northeast (Dakota Co.), east (Burt, Douglas, and Sarpy cos.), southeast (Richardson Co.), and southwest (Frontier and Gosper cos.). The remaining records are from seven counties in east to east-central Nebraska. The oldest remains of the striped skunk from the Holocene of Nebraska appear to be those from Logan Creek site (level B) in Burt County where the material may be as much as 6500 years old. The cranium of a striped skunk from 25DK10 in Dakota Co. was stained red suggesting a possible ceremonial use (O'Shea and Ludwickson, 1992).

Barton and Wisely (2012) reconstructed the Pleistocene and Holocene phylogeography of striped skunks using mitochondrial cytochrome-*b* gene and microsatellite loci. They found that striped skunks were isolated in two areas during the Pleistocene with populations in the eastern United States and the southern United States and Mexico. The southern populations subsequently split into a western population and an intermountain west population. In the Holocene the southern group moved northward into the Great Plains and the intermountain west group moved eastward onto the Great Plains, making a zone of secondary contact after being separated for 150,000 years. Barton and Wisely (2012) stated: "Middle Holocene (~ 8,000 years ago) fossils that appear in Nebraska and Kansas and late Holocene fossils (~5,000 years ago) from several states (i.e. … Nebraska …) support the expansion of striped skunks across the Great Plains throughout the Holocene." The Nebraska fossil data presented here is certainly in line with these conclusions.

***Procyon lotor*—raccoon**

Site records (35).—Brown Co.: 25BW280, 14.5 mi. N, 2 mi. W Johnstown (Pepperl and Falk, 1983: B82). Burt Co.: Logan Creek Site B, C, D, G, 25BT3, Logan Creek, SW Oakland (Snyder and Bozell, 1983: 17; Graham et al., 1987: 283; Widga, 2006: 67-68). Butler Co.: White Site, 25BU20 (Ore, 1981: 11). Cass Co.: Walker Gilmore, 25CC28, 6 mi. SE Murray (Haas, 1983: 252). Dakota Co.: Big Village Site, 25DK5, 1 mi. NE Homer (O'Shea and Ludwickson, 1992: 331). Douglas Co.: Parker Site, 25DO2, on the bluffs north of Ponca Creek and west of the Missouri River, in extreme northeastern Douglas Co. (Ewing, 2000: 4). Franklin Co.: Lost Creek Site, 25FR6, Lost Creek, 2.25 mi. S, 1.5 mi. E Bloomington (Strong, 1935: 95, 100). Frontier Co.: Owens Site, 25FT3, on Medicine Creek (Mick, 1983: 171; Graham et al., 1987: 285); 25FT16, 17, 70, vicinity of Medicine Creek Dam (Mick, 1983: 184, 190, 210; Graham et al., 1987:285-286; Kivett and Metcalf, 1997: 213); 25FT22, ≈ 4 km upstream on Medicine Creek from Medicine Creek Dam (Kivett and Metcalf, 1997: 213); 25FT30, near point where Lime Creek originally entered Medicine Creek (Mick, 1983: 202; Graham et al., 1987: 286; Kivett and Metcalf, 1997: 213); Spring Creek Site, 25FT31, Red Willow Creek, adjacent to Red Willow Dam (Widga, 2004: 29); 25FT39, ≈ 5 km upstream on Medicine Creek from Medicine Creek Dam (Mick, 1983: 210; Graham et al., 1987: 286); Lime Creek Site, 25FT41, north side of Lime Creek valley, 1 mi. from original junction of Lime and Medicine creeks (Davis, 1962: 23; Wedel, 1986: 69; Graham et al., 1987: 283); Allen Site, 25FT50, ≈ 3 mi. upstream from Medicine Creek Dam (Hudson, 2007: 195). Gosper Co.: Wallace Site, 25GO2, Plum Creek, 5.5 mi. N, 9.5 mi. E Elwood (Winfrey, 1991: 82). Hall

Co.: Hulme Site, 25HL28, 16 mi. W Grand Island (Bozell, 1991: 234). Harlan Co.: 25HN36, 2 mi. S Alma (Adair and Brown, 1987: 153, 584). Howard Co.: Stabaco Site, 25HW16, Loup River, near Cushing (Goedert, 1995: 170). Loup Co.: Gracie Creek Site, 25LP16, 0.5 mi. upstream on Gracie Creek from its confluence with the Calamus River (Falk and Pepperl, 1981: A1 151). Nance Co.: 25NC10, vicinity of Fullerton (Mick, 1983: 162; Graham et al., 1987:285). Red Willow Co.: Coon Creek Site, 25RW24, west edge Indianola (Grange, 1980: 167); Doyle Site, 25RW28, 9.5 mi. N, 0.75 mi. W McCook (Grange, 1980: 167, 173). Richardson Co.: Leary Village Site, 25RH1, 4 mi. SE Rulo (Hill and Wedel, 1936: 63). Sarpy Co.: Lucien Fontenelle's Post, 25SY26, Bellevue (Bozell et al., 1990: 29); Patterson Site, 25SY31, ≈ 7 mi. S Gretna (Bozell and Ludwickson, 1999: 83). Sioux Co.: 25SX163, Agate Fossil Beds National Monument, 19.25 mi. S, 7.25 mi. E Harrison (Bozell, 1994a: 49, 62). Washington Co.: Fort Atkinson Site, 25WN9, 0.5 mi. E Fort Calhoun (Carlson, 1979: 124). Webster Co.: Hill Site, 25WT1, 2 mi. S, 7 mi. E Red Cloud (Metcalf, 1941b: 42); Shipman Site, 25WT7, south side of Republican River, between Red Cloud and Guide Rock (Mick, 1983: 154; Graham et al., 1987: 286).

Remarks.—The raccoon is statewide in its distribution in Nebraska (Jones, 1964; Genoways et al., 2008b). It is among the most adaptable species of carnivore and its numbers have increased as human activities have presented new opportunities, such as corn as a food resource (Landholt and Genoways, 2000).

The Holocene records also indicate that the raccoon was statewide in its distribution with remains from the extreme parts of the state—northwest, Sioux Co.; southeast, Richardson Co.; northeast, Dakota Co.; and southwest, Frontier Co. Raccoons (closely followed by the badger) are known from more Holocene sites than any other carnivore in Nebraska. The Holocene record for the raccoon in Nebraska is nearly continuous, beginning with the Allen site (9500-13,000 years ago); Lime Creek site (8000-9500 years ago); Logan Creek site (5340-7070 years ago); Spring Creek site (5500-5850 years ago); 25SX163 (2050-3000 years ago); Doyle site (1420-2160 years ago); White site (1100-2100 years ago); Wallace site (1450-1675 years ago); 25FT70 (650-1410 years ago); Walker Gilmore site (815-1300 years ago). Remains from all other sites are less than 1100 years old. Cullingham et al. (2008) found that raccoons in eastern United States originated from three Pleistocene refugia. The populations west of the Mississippi represent one of these three lineages. The Nebraska Holocene record would indicate that this refugium may have included parts of Nebraska because raccoons appeared early in the record and occurred throughout the period.

ORDER PERISSODACTYLA

Equus caballus—domestic horse

Site records (11).-- Brown Co.: 25BW280, 14.5 mi. N, 2 mi. W Johnstown (Pepperl and Falk, 1983: B82). Butler Co.: Linwood Site, 25BU1, Skull Creek, 1 mi. SW Linwood (Carlson, 1973: 79); Barcal Site, 25BU4, near Abie (O'Shea and Ludwickson, 1992: 338). Dakota Co.: Big Village Site, 25DK5, 1 mi. NE Homer (Steinacher and Carlson, 1984: 34-35; O'Shea and Ludwickson, 1992: 331). Harlan Co.: Stevenson Village, 25HN31, 1.5 mi. S, 0.25 mi. W Alma (Adair and Brown, 1987: 123, 574); White Cat Village, 25HN37, Prairie Dog Creek, 6 mi. SE Alma (Adair and Brown, 1987: 168, 588). Loup Co.: Royal Goodenow Site, 25LP8, confluence of Dry Creek and Calamus River, 13 km NW Burwell (Falk and Pepperl, 1981: A1-80); Gracie Creek Site, 25LP16, 0.5 mi. upstream on Gracie Creek from its confluence with the Calamus River (Falk and Pepperl, 1981: A1 151). Red Willow Co.: 25RW23, 3 mi. N, 6.25 mi. W Indianola; age—early historic (Grange, 1980:1167). Washington Co.: Fort Atkinson Site, 25WN9, 0.5 mi. E Fort Calhoun (Carlson, 1979: 124). Webster Co.: Hill Site, 2 mi. S, 7 mi. E Red Cloud (Wedel, 1936: 62).

Remarks.—There are 11 sites in Nebraska where the remains of the domestic horse have been reported. The domestic horse was introduced to North America by the Spanish some time after AD 1500 so these sites are among the more recent that have been studied. The Royal

Goodenow Site should be among the oldest sites from which the horse has been reported. Upon viewing the chronology of all Nebraska sites it appears that the domestic horse must have reached the area by AD 1700.

ORDER ARTIODACTYLA

***Alces alces*—moose**

Site record (1).—Cass Co.: precise locality unknown.

Remarks.—Bell and Gilmore (1936: 325) stated: "They [flakers] are usually made of deer or elk antler, but for one a tine of moose antler was used. This is of special interest as it is the first evidence of moose reported from eastern Nebraska." Unfortunately, Bell and Gilmore (1936: 325) do not state the specific site where this artifact was discovered, but it is from one of five sites in Cass Co. that they investigated—Behrnes site, Gates site, Rock Bluffs site, Sheldon site, or Woolsey site. For many years, New and Old World moose have been treated as a single species; however, recently Groves and Grubb (2011) have presented an argument for the recognition of two species. Until more data are gathered, I choose to continue to recognize a single species.

There is no evidence that there has been a resident population of moose in Nebraska, but recent data published by Hoffman et al. (2006) document that wandering individuals of moose do enter the state and remain for several years. This flaker could have been produced from one of these wandering individuals or it could be the result of trade with native people in Minnesota or that region.

***Odocoileus hemionus*—mule deer**

Site records (12).—Douglas Co.: Champe Site, 25DO1, margin of uplands overlooking bottom lands through which flow the Elkhorn and Platte rivers (Hill and Cooper, 1936b: 267); Long's Hill, 25DO26, on Ponca Creek in extreme northeastern corner of county (Gilder, 1908: 67). Frontier Co.: 25FT13, 14, 16, 17, vicinity of Medicine Creek Dam (Kivett and Metcalf, 1997: 213); 25FT39, ≈ 5 km upstream on Medicine Creek from Medicine Creek Dam (Kivett and Metcalf, 1997: 213). Lancaster Co.: Schrader Site, 25LC1, Salt Creek, near Roca, about 9 mi. S Lincoln (Hill and Cooper, 1936a: 244-246). Richardson Co.: Leary Village Site, 25RH1, 4 mi. SE Rulo (Hill and Wedel, 1936: 64). Sarpy Co.: Childs Point District, 25SY0, ≈ 1.5 mi. NW Bellevue (Gilder, 1909: 72-73). Sherman Co.: Sweetwater Site, 0.5 mi. NW Sweetwater [in Buffalo Co.] (Champe, 1936: 268). Sioux Co.: Hudson-Meng Site, 25SX115, 23 mi. NW Crawford (Agenbroad, 1978: 36, 65).

Remarks.—Nebraska is one of the states in the central Great Plains where the geographic ranges of mule deer and white-tailed deer overlap, essentially throughout the state (Hall, 1981; Jacobson, 2003). The mule deer is a species of more open habitats and is easily distinguished from the white-tailed deer based on external characteristics, details of the cranium, and shape of the antlers (Hall, 1981). By the early 1900s, the mule deer was nearly extirpated from the state, but complete protection for 40 years and reintroductions have allowed the species to regain most of it geographic range in the state. Jones (1964; Genoways et al., 2008b) believed that modern populations of the mule deer occurred throughout the state with the exception of five counties in the extreme southeastern corner—Gage, Johnson, Nemaha, Pawnee, and Richardson.

Although there are only 12 Holocene sites in Nebraska where mule deer have been definitely identified, its broad distribution in the Holocene is evident. It is present at a site southeast of Rulo in Richardson Co., where modern populations are believed to be absent. There are also other sites along the eastern edge of the state in Douglas and Sarpy counties. Across the state in the northwest corner of the state, mule deer have been identified at the Hudson-Meng Site. There are five sites in Frontier Co. in southwestern Nebraska where mule deer remains have been found. The only part of the state where Holocene mule deer remains have not been identified is in the northeast.

The oldest Holocene record of a mule deer in Nebraska is from the Hudson-Meng site in Sioux Co. where the material is believed to be about 10,000 years old. The material from all remaining sites is less than 1100 years old with the possible exception of the Leary Village site where it could be 1300 years old.

***Odocoileus virginianus*—white-tailed deer**

Site records (22).—Cass Co.: Theodore Davis Site, 25CC17, 1.5 mi. E Weeping Water (Gradwohl, 1969: 143). Chase Co.: Lovitt Site, 25CH1, 12 mi. W Wauneta (Hill and Metcalf, 1942: 204). Dakota Co.: Big Village Site, 25DK5, 1 mi. NE Homer (O'Shea and Ludwickson, 1992: 332). Franklin Co.: Lost Creek Site, 25FR6, Lost Creek, 2.25 mi. S, 1.5 mi. E Bloomington (Strong, 1935: 100). Frontier Co.: 25FT20, 30, near point where Lime Creek originally entered Medicine Creek (Kivett and Metcalf, 1997: 213); 25FT22, ≈ 4 km upstream on Medicine Creek from Medicine Creek Dam (Kivett and Metcalf, 1997: 213; Nepstad-Thornberry et al., 2002: 199); Mowry Bluff Site, 25FT35, 6 mi. NW Cambridge (Falk, 1969b: 48); 25FT36, ≈ 4 km upstream on Medicine Creek from Medicine Creek Dam (Kivett and Metcalf, 1997: 213); 25FT70, vicinity of Medicine Creek Dam (Kivett and Metcalf, 1997: 213). Harlan Co.: White Cat Village, 25HN37, Prairie Dog Creek, 6 mi. SE Alma (Champe, 1949: 289; Gunnerson, 1960: 177). Merrick Co.: Tahaksu Site, 25MK15, 4.8 mi. N, 1.2 mi. W Palmer (Watson, 1996: 135). Nance Co.: Burkett Site, 25NC1, 4 mi. SW Genoa (Dunlevy, 1936: 200); 25NC8, 10, 12, 13, vicinity of Fullerton (Ludwickson, 1978a: 98; Mick, 1983: 166). Richardson Co.: Leary Village Site, 25RH1, 4 mi. SE Rulo (Hill and Wedel, 1936: 63). Sarpy Co.: Gates Site, probably near modern Capehart Road and 36th Street in southern Bellevue (Strong, 1935:153). Stanton Co.: Stanton Site, 25ST1, 1 mi. E Stanton (Gunnerson, n.d.). Valley Co.: Mira Creek Site, 25VY1, 6 mi. W, 0.75 mi. N North Loup (Hill and Kivett, 1940: 183). Webster Co.: Hill Site, 25WT1, 2 mi. S, 7 mi. E Red Cloud (Wedel, 1936: 62).

Remarks.—The geographic range of the white-tailed deer overlaps that of mule deer statewide in Nebraska (Jones, 1964; Hall, 1981). By the early 1900s, the white-tailed deer was nearly extirpated from the state, but complete protection for 40 years and reintroductions have allowed the species to regain most of it geographic range in the state (Jones, 1964; Genoways et al., 2008b). The white-tailed deer inhabits more closed habitats than the mule deer in Nebraska such as the eastern forest and riparian forests along the watercourses in the state. Its numbers have increased dramatically in recent years as populations work the forest edges with easy access to agriculture crops such as corn and soybeans.

There are only 22 Holocene sites in Nebraska where white-tailed deer have been definitely identified. These minimal data may not give a full picture of the Holocene distribution of the species in the state. The eastern part of the state is well represented by sites, such as those in east-central (Cass and Sarpy cos.), southeastern (Richardson Co.), and northeastern (Dakota and Stanton cos.). There are sites in three counties in central Nebraska (Merrick, Nance, and Valley cos.) and in the south-central part of the state (Franklin, Harlan, and Webster cos.). The remaining two counties (Chase and Frontier cos.) are located in southwestern corner of the state with the Lovitt site in Chase County being the western-most Holocene site for the white-tailed deer. This leaves large portions of the state with no Holocene records of the white-tailed deer, including all of the Panhandle, the Sand Hills, and along the Platte River west of Merrick Co.

Hopefully, in the future better methods of identifying the remains of deer will be found so that this lack of records can be more fully evaluated.

The oldest Holocene material of the white-tailed deer in Nebraska is from the Mira Creek site in Valley Co. where the remains may be as much as 2100 years old. Other sites where materials may be slightly older than most sites are 25FT70 in Frontier Co. and Leary Village site in Richardson Co. where remains may be as old as 1400 and 1900 years old, respectively. Material from the remaining 19 Holocene sites was no more than 1100 years old.

***Odocoileus* sp.—deer**

Site records (147).—Banner Co.: Sheep Mountain Site, 25BN1, 5.5 mi. S McGrew [in Scotts Bluff Co.] (Gillaspie, 1955: 19). Boone Co.: Beaver Creek Site, 25BO23, 2.4 km N Loretto (Koch, 2002: i, 1-4; Koch and Nelson, 2002: 87). Boyd Co.: Lynch Site, 25BD1, 0.75 mi. NW Lynch (Witty, 1962: 121-130). Brown Co.: McIntosh Site, 25BW15, Enders Lake, ≈12 mi. S, 12.5 mi. W Ainsworth (Koch, 1995: 44; 2004: 117). Buffalo Co.: Lady Jayne Site, 25BF145, Sec. 11, T9N, R15W (Bleed, 1973: 46); 25BF153, 0.5 mi. NW Miller (Henning, 1977: 64); 25BF189, 5.5 mi. NW Amherst (Henning, 1977: 84). Burt Co.: Logan Creek Site A, B, C, D, F, 25BT3, Logan Creek, SW Oakland (Snyder and Bozell, 1983: 19-20; Graham et al., 1987: 283; Widga, 2006: 67-68). Butler Co.: Linwood Site, 25BU1, Skull Creek, 1 mi. SW Linwood (Carlson, 1973: 79); Barcal Site, 25BU4, near Abie (O'Shea and Ludwickson, 1992: 338); White Site, 25BU20 (Ore, 1981: 11). Cass Co.: Ashland Site, 25CC1, ≈ 3 mi. E Ashland [in Saunders Co.], Occupation B (Hill and Cooper, 1937: 270); Walker Gilmore, 25CC28, 6 mi. SE Murray (Haas, 1983: 251; Graham et al., 1987: 285); Rock Bluffs Site, 25CC31, 1 mi. S Rock Bluff (Strong, 1935: 137; Bell and Gilmore, 1936: 324-325); Little Pawnee Creek Site, 25CC228, 0.8 mi. N, 3 mi. W South Bend (Ludwickson, 1994:116); Behrnes Site, Sec. 14, T10N, R13E (Bell and Gilmore, 1936: 324-325); Gates Site, 2 mi. W Fort Crook, Bellevue (Bell and Gilmore, 1936: 324-325); North Plattsmouth Site, N of Plattsmouth and S of Platte River (Hill and Cooper, 1937: 315); Sheldon Site, Sec. 21, T10N, R13E (Bell and Gilmore, 1936: 324-325); Woolsey Site, Sec. 5, T10N, R14E (Bell and Gilmore, 1936: 324-325). Cedar Co.: Radke Site, 25CD2, 3 mi. S Wynot (Hill, 1988: 115-123); Wiseman Site, 25CD3, 4.5 mi. NE Wynot (Cooper, 1936: 76); Hill, 1988: 115-123); Ferber Site, 25CD10, near Bow Valley Mill (Ludwickson et al., 1981: 22-23). Chase Co.: Lovitt Site, 25CH1, 12 mi. W Wauneta (Gunnerson, 1960: 212-216). Cherry Co.: 25CE309, 3 mi. S, 0.4 mi. W Sparks (Pepperl and Falk, 1983: B85). Cheyenne Co.: Thurston Site, 25CN11, Lodgepole Creek, 6 mi. E Potter (Jensen, 1973: 167). Colfax Co.: Schuyler Site, 25CX1, along Shell Creek, ≈ 3 mi. NW Schuyler (Dunlevy, 1936: 199; Bozell et al., 1982: 28; Graham et al., 1987: 286). Custer Co.: Dickman Ossuary, 25CU7, Sec. 24, T15N, R22W (Philips, 1960a: 20, 22). Dakota Co.: Bobier Site, 25DK1, 4.5 mi. SE Homer (Frantz, 1963: 87-88); Big Village Site, 25DK5, 1 mi. NE Homer (Steinacher and Carlson, 1984: 34-35; O'Shea and Ludwickson, 1992: 332); Nelson Site, 25DK7, 1 mi. N Homer (Frantz, 1963: 87-88); Hancock Site, 25DK14, 1 mi. SE Homer (Frantz, 1963: 87-88). Dawson Co.: 25DS106, Sec. 26, R10N, R19W (Ludwickson, 1978b: 36); 25DS107, 1.25 mi. W Sumner (Ludwickson, 1978b: 36); 25DS109, 6 mi. SW Overton (Ludwickson, 1978b: 36); 25DS111, Sec. 33, T12N, R20W (Ludwickson, 1978b: 36); 25DS112, Sec. 36, T12N, R20W (Ludwickson, 1978b: 36); 25DS120, Sec. 1, T11N, R20W (Ludwickson, 1978b: 36). Dixon Co.: Peterson Site, 25DX1, near Ponca (Hill, 1988: 115-123); Annie's Site, 25DX30, in vicinity of 2 mi. N, 3 mi. E Newcastle (Giessen, 1988: 133-156; Manz and Blakeslee, 1988: 159-160). Douglas Co.: Parker Site, 25DO2, on the bluffs north of Ponca Creek and west of the Missouri River, in extreme northeastern Douglas Co. (Ewing, 2000: 2); 25DO3, Ponca Creek in northeastern corner of county (Cooper, 1939: 119); Saunders Site, 25DO5, eastern side of Elkhorn River in southwestern corner of county (Strong, 1935: 173); Cabanne's Post, 25DO8, near Dodge Park northeast of Omaha (Bozell et al., 1990: 32); 25DO11, 2.5 mi. W Florence (Steinacher and Carlson, 1984: 37); Ponca Creek District, 2.25 mi. N Florence (Gilder, 1907: 714). Deuel Co.: Neumann Site, 25DU3, Lodgepole Creek, 3.5 mi. SE Chappell (Carlson, 1973: 104). Franklin Co.: Red Cloud 3, 3.5 mi. S, 1.5 mi. E Franklin (Wedel, 1934: 152, 201). Frontier Co.: Owens Site, 25FT3, on Medicine Creek (Mick, 1983: 172; Graham et al., 1987: 285); 25FT13, 14, 16, 17, 70, vicinity of Medicine Creek Dam (Mick, 1983: 175, 182, 185, 191, 210; Graham et al., 1987: 285-286; Nepstad-Thornberry et al., 2002: 199); 25FT18, 30, near point where Lime Creek originally entered Medicine Creek (Mick, 1983: 202; Graham et al., 1987: 286; Kivett and Metcalf, 1997: 212); 25FT31, Red Willow Creek, adjacent to Red Willow Dam (Grange, 1980: 166); 25FT22, SE ¼, Sec. 11, T5N, R26W (Turnmire, 1996: 189-191); Mowry Bluff Site, 25FT35, 6 mi. NW Cambridge (Mick, 1983: 205; Graham et al., 1987: 286: 213; Nepstad-Thornberry et al., 2002: 199); 25FT36, ≈ 4 km upstream on Medicine Creek from

Medicine Creek Dam (Mick, 1983: 206); 25FT39, ≈ 5 km upstream on Medicine Creek from Medicine Creek Dam (Mick, 1983: 210; Graham et al., 1987: 286); Lime Creek Site, 25FT41, north side of Lime Creek valley, 1 mi. from original junction of Lime and Medicine creeks (Davis, 1962: 23; Graham et al., 1987: 283; Wedel, 1986: 69); Allen Site, 25FT50, ≈ 3 mi. upstream from Medicine Creek Dam (Holder and Wike, 1949: 261; Bamforth, 2002: 65; Hudson, 2007: 195); 25FT177, vicinity of Medicine Creek Reservoir (Holen and May, 1989: 6); Medicine Creek 4, 2 mi. S Stockville (Wedel, 1934: 166, 201). Garfield Co.: 25GF13, 1 mi. S, 2.1 mi. W Burwell (Bozell and Pepperl, 1982: 10, 12); 25GF14, 1 mi. S, 1.5 mi. W Burwell (Bozell and Pepperl, 1982: 15). Garden Co.: Ash Hollow Cave Site, 25GD2, 3 mi. SE Lewellen (Champe, 1946: 43). Gosper Co.: Wallace Site, 25GO2, Plum Creek, 5.5 mi. N, 9.5 mi. E Elwood (Winfrey, 1991: 83); Mousel Site (Turkey Creek I), 25GO5, 16 mi. S, 5 mi. W Elwood (Philips, 1957a: 3); Scheivalbein Site, 25GO21, 3 mi. W, 16 mi. S Elwood (Philips, 1954: 1); Golter Site, 25GO30, 4 mi. N Arapahoe [in Furnas Co.] (Philips, 1957b: 1); Stillman Site, 25GO35, about 15 mi. S, 8 mi. W Elwood (Philips, 1960b: 1-2); Graf Site, 25GO44, 13 mi. S, 9 mi. W Elwood (Phillips, 1957c: 2); Gross Site, 25GO53, 12 mi. S, 10 mi. W Elwood (Philips, 1956b: 2); Lowell Dawson Site, 25GO87, Sec. 1, T5N, R22W (Philips, 1959: 2); Dawson Site, 25GO88, Sec. 12, T5N, R22W (Philips, 1960d: 3). Hall Co.: Hulme Site, 25HL28, 16 mi. W Grand Island (Bozell, 1991: 234). Harlan Co.: Stevenson Village, 25HN31, 1.5 mi. S, 0.25 mi. W Alma (Adair and Brown, 1987: 123, 574); 25HN36, 2 mi. S Alma (Adair and Brown, 1987: 153, 584); White Cat Village, 25HN37, Prairie Dog Creek, 6 mi. SE Alma (Adair and Brown, 1987: 168, 588); Green Plum Site, 25HN39, 5.5 mi. S, 0.8 mi. E Alma (Adair and Brown, 1987: 182, 589); 25HN62, 4.5 mi. S, 2 mi. E Alma (Adair and Brown, 1987: 262, 593); 25HN64, 0.75 S, 2.5 mi. W Alma (Moore, 1988: 40, 143); 25HN65, 1.25 mi. W Alma (Moore, 1988: 145); 25HN125, 0.75 mi. S, 2.1 mi. E Alma (Adair and Brown, 1987: 276, 594). Hitchcock Co.: Carmody Site, 25HK7, 2 mi. W Trenton (Kivett, 1951: 57-58; Kivett, 1952: 31); Massacre Canyon Site, 25HK13, ≈ 3 mi. E Trenton (Kivett, 1951: 36, 38; Kivett, 1952: 21-23). Holt Co.: 25HT1, 17 mi. N, 3 mi. W O'Neill (Price, 1956: 81); 25HT3, 0.2 mi. N Redbird (Wood, 1965: 107-108, 112). Hooker Co.: Humphrey Site, 25HO21, Middle Loup River, about 5 mi. E Mullen (Gunnerson, 1960: 204); Kelso Site, 25HO23, Middle Loup River, about 5 mi. E Mullen (Kivett, 1952: 39-40). Howard Co.: Stabaco Site, 25HW16, Loup River, near Cushing (Goedert, 1995: 170); Schmidt Site, 25HW301, along North Loup River, near Elba (Mick, 1983: 157; Graham et al., 1987: 284). Keya Paha Co.: 25KP243, East Middle Creek, 2.25 mi. S, 1.25 mi. E Norden (Pepperl and Falk, 1983: B91). Knox Co.: Ponca Fort, 25KX1, 1 mi. E Verdel (Wood, 1993: 65); Minarik Site, 25KX2, 9, Ponca Creek, 7 mi. NW Verdel (Wood, 1956: 115-116, 120, 138-139; 1965: 107-109, 112); 25KX4, Niobrara River, 14 mi. upstream from confluence of Niobrara and Missouri rivers (Wood, 1956: 80); 25KX15, about 4 mi. N, 19.5 mi. Niobrara (Blakeslee and O'Shea, 1983: 177-179); 25KX56, mouth of Devils Nest Creek, about 3.75 mi. N, 15 mi. E Niobrara (Blakeslee and O'Shea, 1983: 189); 25KX68, about 2 mi. N, 6.25 mi. E Niobrara (Blakeslee and O'Shea, 1983: 216); 25KX200, about 4.5 mi. N, 23 mi. E Niobrara; (Blakeslee and O'Shea, 1983: 233-234); 25KX203, Deepwater Recreation Area, about 4.5 mi. N, 25 mi. E Niobrara (Blakeslee and O'Shea, 1983: 239-242); Tramp Deep Site, 25KX204, about 5 mi. E Niobrara (Howard and Gant, 1966: 18, 22, 38). Loup Co.: Royal Goodenow Site, 25LP8, confluence of Dry Creek and Calamus River, 8 mi. NW Burwell [in Garfield Co.] (Falk and Pepperl, 1981: A1-80-81; Roper, 1989: 313-316); Jane Toben Site, 25LP11, 0.5 mi. N confluence of Dry Creek and Calamus River (Falk and Pepperl, 1981: A1 93-106). Madison Co.: 25MD3, near Union Creek, 1.75 mi, E Madison (Steinacher and Carlson, 1984: 53-55). McPherson Co.: Dry Lake Burial Site, 25MP2, Dry Lake, 13 mi. NW Tryon (Oothoudt, 1976: 102-109). Morrill Co.: Greenwood Site, old Greenwood Stage Station, Keenan Ranch, ≈ 9 mi. S, 8 mi. E Redington (Renaud, 1933: 14). Nance Co.: Burkett Site, 25NC1, 4 mi. SW Genoa (Blackman, 1907: 339, 341); 25NC8, 10, vicinity of Fullerton (Mick, 1983: 160, 163; Graham et al., 1987:285). Nemaha Co.: Leahy Site, 25NH1, ≈ 3 mi. W Peru (Hill and Kivett, 1940: 199); Majors Site, 25NH2, ≈ 3 mi. NW Peru (Hill and Cooper, 1937: 324); Osborne Site, 25NH5, Missouri River, N of Brownsville (Hill and Cooper, 1937: 338). Nuckolls Co.: 25NO2, Sec. 3, T4N, R7W, terrace about 0.25 mi. Blue River (Frantz, 1965: 60). Otoe Co.: 25OT24, Sec. 26, T8N, R14E (Kaldahl, 1993: 106-112). Pawnee Co.: Table Rock Site, 25PW1, Sec. 33, T3N, R12E (Bell and Gilmore, 1936: 324-325). Platte Co.: Hill-Rupp Site, 25PT13, 1.5 mi. N Monroe (Metcalf, 1941a: 34). Red Willow Co.: 25RW22, 3 mi. N, 6.25 mi. W Indianola (Grange 1980: 166; Mick, 1983: 169); Doyle Site, 25RW28, 9.5 mi. N, 0.75 mi. W McCook (Grange, 1980: 167); Gillen Pits, 25RW101, 4.5 mi. W McCook (Corner, 1977: 79, 84); Davidson Pits, 25RW102, 1 mi. W Bartley (Corner, 1977: 79-80). Richardson Co.: Rulo Southeast, 25RH69, 0.4 km S Rulo (Nelson, 2006: 165); Rulo Southeast, 25RH70, 0.6 km S Rulo (Nelson, 2006: 170, 173). Sarpy Co.: Childs Point District, 25SY0, ≈ 1.5 mi. NW Bellevue (Gilder, 1909: 72); Fremont 1, 25SY1, margin of uplands overlooking bottom lands through which flow the Elkhorn and Platte rivers (Hill and Cooper, 1936c: 287-288); Fontenelle Forest Site, 25SY3, near Bellevue (Hill and Cooper, 1937: 309); Lucien Fontenelle's Post, 25SY26, Bellevue (Bozell et al., 1990: 30); Patterson Site, 25SY31, ≈ 7 mi. S Gretna (Bozell and Ludwickson, 1999: 84); unnamed site, near Meadow (Brooking, 1921: 38). Scotts Bluff Co.: Signal Butte Site, 25SF1, Signal Butte, 11.5 mi. S Morrill (Strong, 1935: 235); Gering Burial Site, 25SF10, 1.5 mi.

NW Melbeta (Oothoudt, 1976: 68-72); 25SF76, Scotts Bluff National Monument, 0.7 mi. N, 0.1 mi. W Visitor Center (Bozell, 1994b: 52, 63). Sherman Co.: Bill Packer Site, 25SM9, along Davis Creek in extreme northeastern corner of county (Graham et al., 1987: 285; Bozell and Rogers, 1989: 28). Sioux Co.: 25SX163, Agate Fossil Beds National Monument, 19.25 mi. S, 7.25 mi. E Harrison (Bozell, 1994a: 49, 62). Stanton Co.: Stanton Site, 25ST1, 1 mi. E Stanton (Gunnerson, n.d.). Thurston Co.: Gerald Rose Site, 25TS10, vicinity of Winnebago (Gant, 1966: 6; Carlson, 1973: 53-54). Washington Co.: Fort Atkinson Site, 25WN9, 0.5 mi. E Fort Calhoun (Carlson, 1979: 124); Site W1, on Douglas-Washington Co. line in extreme southeastern corner of county (Cooper, 1939: 111). Webster Co.: Hill Site, 25WT1, 2 mi. S, 7 mi. E Red Cloud (Metcalf, 1941b: 30); Shipman Site, 25WT7, south side of Republican River, between Red Cloud and Guide Rock (Mick, 1983: 154; Graham et al., 1987: 286).

Remarks.—Unfortunately, it is extremely difficult, if not impossible, to distinguish the two species of deer occurring in Nebraska based on post-cranial skeletons, partial cranial material, and even on some incomplete antlers. Jacobson (2003) has made an attempt to find characters in the post-cranial skeletons to separate the species, but many of the characters involve judgment decisions and require extensive comparative material. Also the condition of the archeological material can heavily influence the ability to identify the species. This situation is unfortunate because a great deal of information is lost that would be of interest to mammalogists, archeologists, and anthropologists.

Two sites yielded unusually large concentrations of deer remains. At the Hulme Site in Hall Co., which is about 800 years old, Bozell (1991: 234) found the minimum number of individuals represented to be 148. The excavations at Lucien Fontenelle's Post, near Bellevue, Sarpy Co. (Bozell et al., 1990: 30) yielded a minimum of 399 individuals. The deer remains at this trading post, which was operated between 1822 and 1842, probably were the results of market hunting for use at the post.

***Cervus canadensis*—elk or wapiti**

Site records (58).—Boyd Co.: Lynch Site, 25BD1, 0.75 mi. NW Lynch (Witty, 1962: 126, 131). Buffalo Co.: 25BF193, 0.75 mi. W Amherst (Henning, 1977: 86). Burt Co.: Logan Creek Site B, C, F, 25BT3, Logan Creek, SW Oakland (Snyder and Bozell, 1983: 19; Graham et al., 1987: 283; Widga, 2006: 67-68). Butler Co.: Barcal Site, 25BU4, near Abie (O'Shea and Ludwickson, 1992: 338). Cass Co.: Ashland Site, 25CC1, ≈ 3 mi. E Ashland [in Saunders Co.], Occupation A (Hill and Cooper, 1937: 263); Theodore Davis Site, 25CC17, 1.5 mi. E Weeping Water (Gradwohl, 1969: 143); Walker Gilmore, 25CC28, 6 mi. SE Murray (Strong, 1935: 192). Cedar Co.: Schulte Site, 25CD1, St. Helena (Cooper, 1936: 76); Wiseman Site, 25CD3, 4.5 mi. NE Wynot (Cooper, 1936: 76); Ferber Site, 25CD10, near Bow Valley Mill (Ludwickson et al., 1981: 22-23). Chase Co.: Lovitt Site, 25CH1, 12 mi. W Wauneta (Gunnerson, 1960: 212-216). Colfax Co.: Schuyler Site, 25CX1, along Shell Creek, ≈ 3 mi. NW Schuyler (Bozell et al., 1982: 28; Graham et al., 1987: 286). Dakota Co.: Bobier Site, 25DK1, 4.5 mi. SE Homer (Frantz, 1963: 104); Big Village Site, 25DK5, 1 mi. NE Homer (Steinacher and Carlson, 1984: 34-35; O'Shea and Ludwickson, 1992: 332). Dixon Co.: Peterson Site, 25DX1, near Ponca (Hill, 1988: 110, 120). Douglas Co.: Saunders Site, 25DO5, eastern side of Elkhorn River in southwestern corner of county (Strong, 1935: 173); Cabanne's Post, 25DO8, near Dodge Park northeast of Omaha (Bozell et al., 1990: 32); Ponca Creek District, 2.25 mi. N Florence (Gilder, 1907: 706, 714). Franklin Co.: Lost Creek Site, 25FR6, Lost Creek, 2.25 mi. S, 1.5 mi. E Bloomington (Strong, 1935: 97-100). Frontier Co.: 25FT17, vicinity of Medicine Creek Dam (Kivett and Metcalf, 1997: 213); Lime Creek Site, 25FT41, north side of Lime Creek valley, 1 mi. from original junction of Lime and Medicine creeks (Davis, 1962: 23; Wedel, 1986: 69). Garden Co.: Ash Hollow Cave Site, 25GD2, 3 mi. SE Lewellen (Champe, 1946: 43). Hall Co.: Hulme Site, 25HL28, 16 mi. W Grand Island (Bozell, 1991: 234). Harlan Co.: White Cat Village, 25HN37, Prairie Dog Creek, 6 mi. SE Alma (Adair and Brown, 1987: 168, 588); Blue Stone Site, 25HN45, Prairie Dog Creek, 7 mi. SE Alma (Rusco, 1960: 40). Hooker Co.: Humphrey Site, 25HO21, Middle Loup River, about 5 mi. E Mullen (Gunnerson, 1960: 204). Holt Co.: 25HT1, 17 mi. N, 3 mi. W O'Neill (Price, 1956: 81). Howard Co.: unnamed site, Munson Creek, 3 mi. W Elba (Hill, 1932: 172-175). Knox Co.: Ponca Fort, 25KX1, 1 mi. E Verdel (Wood, 1993: 42, 64); Minarik Site, 25KX2, 25KX9, Ponca Creek, 7 mi. NW Verdel (Wood, 1956: 138; 1965: 109-112); 25KX200, about 4 1/2 mi. N, 23 mi. E Niobrara; (Blakeslee and O'Shea, 1983: 233-234); Niobrara Railroad Bridge

Site, 25KX207, confluence of Niobrara and Missouri rivers (Howard and Gant, 1966: 25, 38). Loup Co.: Royal Goodenow Site, 25LP8, confluence of Dry Creek and Calamus River, 13 km NW Burwell (Roper, 1989: 313-316). Madison Co.: 25MD3, near Union Creek, 1.75 mi. E Madison (Steinacher and Carlson, 1984: 53-55). Merrick Co.: Tahaksu Site (25MK15), 4.8 mi. N, 1.2 mi. W Palmer (Watson, 1996: 135). Nance Co.: Burkett Site, 25NC1, 4 mi. SW Genoa (Dunlevy, 1936: 200; Wedel, 1936: 41); 25NC8, 10, 12, 13, vicinity of Fullerton (Ludwickson, 1978a: 98; Mick, 1983: 159, 162, 165; Graham et al., 1987:285); unnamed site, 3 mi. E Genoa (Wedel, 1936: 54). Platte Co.: Bakenhus ossuary, 25PT4, Sec. 15, T19N, R1E (Lamb, 1939b: 11); Feye Site, 25PT9, 5.5 mi. S Creston (Kivett, 1952: 57); Hill-Rupp Site, 25PT13, 1.5 mi. N Monroe (Metcalf, 1941a: 24, 34). Richardson Co.: Leary Village Site, 25RH1, 4 mi. SE Rulo (Hill and Wedel, 1936: 64); Rulo Southeast, 25RH69, 0.4 km S Rulo (Nelson, 2006: 165); Rulo Southeast, 25RH70, 0.6 km S Rulo (Nelson, 2006: 170, 173). Sarpy Co.: Childs Point District, 25SY0, ≈ 1.5 mi. NW Bellevue (Gilder, 1909: 71, 73); Gates Site, 25SY5, probably near modern Capehart Road and 36th Street in southern Bellevue (Strong, 1935:146); Lucien Fontenelle's Post, 25SY26, Bellevue (Bozell et al., 1990: 30); unnamed site, near Meadow (Brooking, 1921: 38-39). Scotts Bluff Co.: Signal Butte Site, 25SF1, Signal Butte, 11.5 mi. S Morrill (Strong, 1935: 234). Sherman Co.: Sweetwater Site, 0.5 mi. NW Sweetwater [in Buffalo Co.] (Champe, 1936: 268-269). Stanton Co.: Stanton Site, 25ST1, 1 mi. E Stanton (Gunnerson, n.d.). Webster Co.: Shipman Site, 25WT7, south side of Republican River, between Red Cloud and Guide Rock (Mick, 1983: 154; Graham et al., 1987: 286).

Remarks.—The elk was statewide in its historic distribution in Nebraska, but by the 1880s the species had been extirpated from the state (Jones, 1964; Genoways et al., 2008b); however, by the 1950s individuals began appearing in the northwestern part of the state. An estimated 1400 elk were in the state occupying areas in western and central Nebraska by 2008 (Fricke et al., 2008).

This species has been known over the last 40 years by the scientific name *Cervus elaphus*. However, recent work by Polziehn and Strobeck (2002) and Ludt et al. (2004) has provided evidence that we should revert to the earlier name *C. canadensis* for this species. Their genetic studies have shown that the red deer of Europe (*C. elaphus*) is a species distinct from the elk of Asia and North America to which the name C. *canadensis* should be applied.

Some of these records were recently presented by Fricke et al. (2008) in a discussion of the historic distribution of elk in Nebraska. The more complete record of the Holocene elk in Nebraska presented here can update our understanding of this record. There are 56 sites from 31 counties where Holocene elk remains have been found in the state. These sites support the idea of elk having a statewide distribution extending well back into the Holocene. In addition, plotting the sites indicates a close association between the elk sites and the major river systems in Nebraska—Missouri, Platte, North Platte, Niobrara, Republican, Elkhorn, and Loup. Some of this association may be an artifact resulting from where the Native Americans had placed their camps, but I believe that much of this association is real. Elk are large, highly mobile herbivores that are capable of covering long distances. Historic accounts indicate that elk were certainly observed in various situations, including on the open plains (Fricke et al., 2008), but it does seem more than reasonable that they would have taken advantage of the cover provided along the river systems, including trees, tall grass and forbs, and associated ravines and draws.

The Holocene records for elk in Nebraska do not cover the entire time span of the epoch. The earliest record is from the Lime Creek site in Frontier Co. where the remains fall into the time period of 8000 to 9500 years ago. The next oldest sites are at opposite ends of the state—in eastern Nebraska the oldest site is from as much as 7500 years ago from Logan Creek site, level F, in Burt Co. and in western Nebraska the next oldest site is Signal Butte in Scotts Bluff Co., where the elk remains may be as much as 5000 years old.

Antilocapra americana—pronghorn

Site records (72).—Boone Co.: Beaver Creek Site, 25BO23, 2.4 km N Loretto (Koch, 2002: i, 1-4; Koch and Nelson, 2002: 87). Brown Co.: McIntosh Site, 25BW15, Enders Lake, ≈12 mi. S, 12.5 mi. W Ainsworth (Koch, 1995: 44; 2004: 117); 25BW252, 16 mi. N, 5.5 mi. W Johnstown (Pepperl and Falk, 1983: B81). Buffalo Co.: 25BF161, Sec. 18, T10N, R17W (Ludwickson, 1978b: 35); 25BF187, 3.75 mi. NW Amherst (Ludwickson, 1978b: 35). Burt Co.: Logan Creek Site B, C, D, 25BT3, Logan Creek, SW Oakland (Snyder and Bozell, 1983: 20; Graham et al., 1987: 283; Widga, 2006: 67-68). Butler Co.: Barcal Site, 25BU4, near Abie (O'Shea and Ludwickson, 1992: 338). Cass Co.: Walker Gilmore, 25CC28, 6 mi. SE Murray (Strong, 1935: 194). Cedar Co.: Ferber Site, 25CD10, near Bow Valley Mill (Ludwickson et al., 1981: 22-23). Chase Co.: Lovitt Site, 25CH1, 12 mi. W Wauneta (Hill and Metcalf, 1942: 204; Gunnerson, 1960: 212-216). Cherry Co.: 25CE309, 3 mi. S, 0.4 mi. W Sparks (Pepperl and Falk, 1983: B85). Cheyenne Co.: Thurston Site, 25CN11, Lodgepole Creek, 6 mi. E Potter (Jensen, 1973: 167). Colfax Co.: Schuyler Site, 25CX1, along Shell Creek, ≈ 3 mi. NW Schuyler (Bozell et al., 1982: 28; Graham et al., 1987: 286). Dakota Co.: Hancock Site, 25DK14, 1 mi. SE Homer (Frantz, 1963: 97-98). Dawes Co.: 25DW59, 0.7 mi. W Crawford (Bozell and Ludwickson, 1988: 86). Dawson Co.: 25DS118, Sec. 16, T10N, R19W (Ludwickson, 1978b: 36). Douglas Co.: Ponca Creek District, 2.25 mi. N Florence (Gilder, 1907: 706, 711). Deuel Co.: Neumann Site, 25DU3, Lodgepole Creek, 3.5 mi. SE Chappell (Carlson, 1973: 104). Franklin Co.: Lost Creek Site, 25FR6, Lost Creek, 2.25 mi. S, 1.5 mi. E Bloomington (Strong, 1935: 100). Frontier Co.: Owens Site, 25FT3, on Medicine Creek (Mick, 1983: 172; Graham et al., 1987: 285); 25FT13, 14, 16, 17, 30, vicinity of Medicine Creek Dam (Mick, 1983: 175, 182, 185, 191, 202; Graham et al., 1987: 285-286; Kivett and Metcalf, 1997: 213); 25FT18, 20, near point where Lime Creek originally entered Medicine Creek (Kivett and Metcalf, 1997: 212-213); 25FT22, 36, ≈ 4 km upstream on Medicine Creek from Medicine Creek Dam (Turnmire, 1996: 189-191; Nepstad-Thornberry et al., 2002: 199; Mick, 1983: 206; Kivett and Metcalf, 1997: 213); Spring Creek Site, 25FT31, Red Willow Creek, adjacent to Red Willow Dam (Grange, 1980: 166; Widga, 2004: 29); Mowry Bluff Site, 25FT35, 6 mi. NW Cambridge [in Furnas Co.] (Falk, 1969a: 42, 1969b: 48; Mick, 1983: 205; Graham et al., 1987: 286; Nepstad-Thornberry et al., 2002: 199); 25FT39, ≈ 5 km upstream on Medicine Creek from Medicine Creek Dam (Mick, 1983: 210; Graham et al., 1987: 286; Kivett and Metcalf, 1997: 213); Lime Creek Site, 25FT41, north side of Lime Creek valley, 1 mi. from original junction of Lime and Medicine creeks (Davis, 1962: 23; Graham et al., 1987: 283; Wedel, 1986: 69); Allen Site, 25FT50, ≈ 3 mi. upstream from Medicine Creek Dam (Holder and Wike, 1949: 261; Bamforth, 2002: 65; Hudson, 2007: 195); 25FT54, Red Willow Creek, 5 mi. upstream from Red Willow Dam (Grange, 1980: 166; Mick, 1983: 169); 25FT70, vicinity of Medicine Creek Dam (Mick, 1983: 210; Graham et al., 1987: 286; Kivett and Metcalf, 1997: 213; Nepstad-Thornberry et al., 2002: 199). Garden Co.: Ash Hollow Cave Site, 25GD2, 3 mi. SE Lewellen (Champe, 1946: 43). Gosper Co.: Wallace Site, 25GO2, Plum Creek, 5.5 mi. N, 9.5 mi. E Elwood (Winfrey, 1991: 83). Hall Co.: Hulme Site, 25HL28, 16 mi. W Grand Island (Bozell, 1991: 234). Harlan Co.: 25HN36, 2 mi. S Alma (Adair and Brown, 1987: 154, 585); 25HN40, 0.8 mi. S, 2 mi. E Alma (Adair and Brown, 1987: 194, 590). Hooker Co.: Humphrey Site, 25HO21, Middle Loup River, about 5 mi. E Mullen (Gunnerson, 1960: 204); Kelso Site, 25HO23, Middle Loup River, about 5 mi. E Mullen (Kivett, 1952: 39-40). Howard Co.: Schmidt Site, 25HW301, along North Loup River, near Elba (Mick, 1983: 157; Graham et al., 1987: 284). Merrick Co.: Tahaksu Site, 25MK15, 4.8 mi. N, 1.2 mi. W Palmer (Watson, 1996: 135). Morrill Co.: Greenwood Site, old Greenwood Stage Station, Keenan Ranch, ≈ 9 mi. S, 8 mi. E Redington (Renaud, 1933: 14). Nance Co.: 25NC8, 10, 13 vicinity of Fullerton (Ludwickson, 1978a: 98; Mick, 1983: 160, 163, 166; Graham et al., 1987:285); Palmer Locality [in part], 25NC29, 5.5 mi. S, 19 mi. W Fullerton (Meadow and Peterson, 2001: 157). Platte Co.: Hill-Rupp Site, 25PT13, 1.5 mi. N Monroe (Metcalf, 1941a: 34). Red Willow Co.: 25RW22, 3 mi. N, 6.25 mi. W Indianola (Grange 1980: 166; Mick, 1983: 169); Doyle Site, 25RW28, 9.5 mi. N, 0.75 mi. W McCook (Grange, 1980: 167); Gillen Pits, 25RW101, 4.5 mi. W McCook (Corner, 1977: 79, 85-86); Davidson Pits, 25RW102, 1 mi. W Bartley (Corner, 1977: 79, 85-86). Sarpy Co.: Childs Point District, 25SY0, ≈ 1.5 mi. NW Bellevue (Gilder, 1909: 72-73); Lucien Fontenelle's Post, 25SY26, Bellevue (Bozell et al, 1990: 30); Patterson Site, 25SY31, ≈ 7 mi. S Gretna (Bozell and Ludwickson, 1999: 84). Scotts Bluff Co.: Signal Butte Site, 25SF1, 11.5 mi. S Morrill (Strong, 1935: 236). Sherman Co.: Bill Packer Site, 25SM9, along Davis Creek in extreme northeastern corner of county (Graham et al., 1987: 285; Bozell and Rogers, 1989: 28); Sweetwater Site, 0.5 mi. NW Sweetwater [in Buffalo Co.] (Champe, 1936: 268). Sioux Co.: Hudson-Meng Site, 25SX115, 23 mi. NW Crawford (Agenbroad, 1978: 36); 25SX157, Agate Fossil Beds National Monument, 19.25 mi. S, 5.75 mi. E Harrison (Bozell, 1993: 61; Bozell, 1994a: 48, 62); 25SX163, Agate Fossil Beds National Monument, 19.25 mi. S, 7.25 mi. E Harrison (Bozell, 1993: 64); 25SX476, Agate Fossil Beds National Monument, 1.7 mi. E Agate (Bozell, 1993: 67); 25SX486, Agate Fossil Beds National Monument, 19.25 mi. S, 5.25 mi. E Harrison (Bozell, 1994a: 50, 63); 25SX487, Agate Fossil Beds National Monument, 0.1 mi. N, 0.15 mi. E Visitor Center (Bozell, 1993: 67). Stanton Co.: Stanton Site, 25ST1, 1 mi. E Stanton (Gunnerson, n.d.). Webster Co.: Hill Site, 2 mi. S, 7 mi. E Red Cloud (Wedel, 1936: 62); Shipman

Site, 25WT7, south side of Republican River, between Red Cloud and Guide Rock (Mick, 1983: 154; Graham et al., 1987: 286).

Remarks.—Jones (1964; Genoways et al., 2008b) mapped the historic geographic range of the pronghorn as occurring throughout Nebraska with the possible exception of the eastern portions of the counties bordering on the Missouri River. Walker (2000) also presented Holocene localities in Nebraska for *A. americana* based on FAUNMAP data, but only plotted them on the maps without recording specific localities. By the early 20th century the pronghorn was either extirpated from the state or represented by a few scattered individuals in the western-most portion of the Panhandle (Jones, 1964; Hoffman et al., 2011). Nebraska gave pronghorns complete protection beginning in 1907. From this point, numbers slowly began to rebound through natural migration from the west and southwest and translocation of individuals. Numbers reached 187 by 1925 and by 1953 a limited hunting season was allowed. Pronghorn populations currently occur throughout the Panhandle and Sand Hills and a small area in the southwestern corner of the state (Hoffman et al., 2011).

Hoffman et al. (2011) have presented and discussed all of the Holocene sites listed here. There are 72 pronghorn Holocene sites in 35 Nebraska counties. Hoffman et al. (2011) found that: "There is an almost continuous record in space and time through the Holocene." The oldest records of pronghorn fossils in Nebraska are from the earliest Holocene-Pleistocene transition sites in Red Willow Co.—Gillen Pits and Davidson Pits. Pronghorn are represented in several of the earliest archeological sites in the state—Allen site and Limestone Creek site, Frontier Co. (as much as 13,000 years ago and 9500 years ago, respectively); Hudson-Meng, Sioux Co. (10,000 years ago); and Logan Creek site, Burt Co. (7000 years ago). There is another group of Holocene sites that fall within the Plains Archaic period covering the period of 1500 to 8000 years ago—25DW59, Dawes Co.; Spring Creek site, Frontier Co.; Signal Butte site, Scotts Bluff Co.; and 25SX157 and 25SX163, Sioux Co. "All other pronghorn remains in Nebraska come from sites that are less than 2000 years old" (Hoffman et al., 2011).

Nebraska is along the eastern edge of the distribution of pronghorns. In the historic record (Hoffman et al., 2011), it appears that the pronghorn may not have occurred in the eastern-most counties of Nebraska along the Missouri River. In the Holocene record there are seven sites in five counties (Burt, Cass, Dakota, Douglas, and Sarpy) along the Missouri River that contain pronghorn remains. An inspection of this pronghorn material reveals that at five sites these remains have been modified for utilitarian or ceremonial purposes. Only at Logan Creek site (25BT3) and Walker Gilmore site (25CC28) did the pronghorn remains appear to be unaltered. These observations would tend to support the idea that pronghorns did not occur in any large number along the Missouri River even in the Holocene. Hoffman et al. (2011) concluded that the eastern distribution of the pronghorn was limited by the forest along the Missouri River as Jones (1964) had suggested, but also, and more importantly, the tall grass prairie occupying much of this part of the state.

At most Holocene sites the pronghorn skeletal material is only a small portion of the remains recovered (Hoffman et al., 2011). At the majority of sites, among the game animals, bison remains outnumbered all others by far. Following bison are usually the remains of deer, elk, and in fourth place pronghorn. However, there are at least two exceptions to this pattern—Lime Creek site (25FT41), Frontier Co. and Hulme site (25HL28), Hall Co. In the oldest zone at the Lime Creek site (about 9500 years ago) beaver and pronghorn bones predominate, whereas in younger layers bison bones become dominant. Pronghorn bones are the most abundant followed by deer in the Hulme site (AD 1170 to AD 1220) with a minimum of 306 individual pronghorns represented.

Bison bison—American bison

Site records (276).—Antelope Co.: Rosedale Reservoir, 25AP11, west-central Antelope Co., southwest of Clearwater (Gunnerson and Gunnerson, 1952: 21); 25AP37, northwestern edge of Neligh (Buhta et al., 2007: 138). Banner Co.: Sheep Mountain Site, 25BN1, 5.5 mi. S McGrew [in Scotts Bluff Co.] (Gillaspie, 1955: 18-19); 25BN28, Pumpkin Creek, 7 mi. NE Harrisburg (Steinacher and Carlson, 1984: 2). Boone Co.: Beaver Creek Site, 25BO23, 2.4 km N Loretto (Koch, 2002: i, 1-4; Koch and Nelson, 2002: 87). Boyd Co.: Lynch Site, 25BD1, 0.75 mi. NW Lynch (Freed, 1954: 44-48; Witty, 1962: 119-130); 25BD124, north end Sunshine Bottoms (Teter, 1975: 90-91, 143); Ponca Creek Reservoir, 25BD201, 2.5 mi. NW Butte (White and Cooper, 1951: 29-30); Ponca Creek, 3.5 mi. NW Butte (White, 1951: 29-30). Brown Co.: McIntosh Site, 25BW15, Enders Lake, ≈12 mi. S, 12.5 mi. W Ainsworth (Koch, 1995: 44; 2004: 117); 25BW252, 16 mi. N, 5.5 mi. W Johnstown (Pepperl and Falk, 1983: B81). Buffalo Co.: 25BF2, 0.5 mi. NW Amherst (Ludwickson, 1978b: 34); Post Draw Site, 25BF101, Sec. 35, T11N, R13W (Ludwickson, 1978b: 34); 25BF105, Sec. 10, T10N, R13W (Ludwickson, 1978b: 34); 25BF109, Sec. 4, T9N, R16W (Henning, 1977: 30; Ludwickson, 1978b: 34); 25BF139, Sec. 10, T10N, R13W (Ludwickson, 1978b: 34); 25BF140, Sec. 35, T11N, R13W (Ludwickson, 1978b: 34); 25BF142, Sec. 16, T9N, R15W (Ludwickson, 1978b: 34); Lady Jayne Site, 25BF145, Sec. 11, T9N, R15W (Bleed, 1973: 46); 25BF148, Sec. 7, T9N, R14W (Ludwickson, 1978b: 34); 25BF153, 0.5 mi. NW Miller (Henning, 1977: 64; Ludwickson, 1978b: 34); 25BF155, 0.125 mi. N Miller (Ludwickson, 1978b: 35); 25BF159, 2.25 mi. NW Amherst (Ludwickson, 1978b: 35); 25BF161, Sec. 18, T10N, R17W (Ludwickson, 1978b: 35); 25BF178, Sec. 6, T10N, R17W (Ludwickson, 1978b: 35); 25BF189, 5.5 mi. NW Amherst (Henning, 1977: 84; Ludwickson, 1978b: 35); 25BF193, 0.75 mi. W Amherst (Ludwickson, 1978b: 35); 25BF199, 4.5 mi. NW Amherst (Henning, 1977: 89; Ludwickson, 1978b: 35); Flat Rock Site, 25BF210, Sec. 31, T10N, R16W (Bleed, 1973: 98-99). Burt Co.: Logan Creek Site A, B, C, D, F, G, 25BT3, Logan Creek, SW Oakland (Snyder and Bozell, 1983: 20; Graham et al., 1987: 283; Widga, 2006: 67-68); 25BT11, 3.5 mi. E Bancroft (Carlson, 1973:10). Butler Co.: Linwood Site, 25BU1, Skull Creek, 1 mi. SW Linwood (Carlson, 1973: 79); Barcal Site, 25BU4, near Abie (O'Shea and Ludwickson, 1992: 338); White Site, 25BU20 (Ore, 1981: 11). Cass Co.: Ashland Site, 25CC1, ≈ 3 mi. E Ashland [in Saunders Co.], Occupations A and C (Hill and Cooper, 1937: 263, 273); Ashland Site, 25CC1, ≈ 3 mi. E Ashland [in Saunders Co.], Occupations B (Hill and Cooper, 1937: 270); Theodore Davis Site, 25CC17, 1.5 mi. E Weeping Water (Gradwohl, 1969: 143); Walker Gilmore, 25CC28, 6 mi. SE Murray (Sterns, 1915: 125; Haas, 1983: 252; Graham et al., 1987: 285); Rock Bluffs Site, 25CC31, 1 mi. S Rock Bluff (Strong, 1935: 137; Bell and Gilmore, 1936: 324-325); Behrnes Site, Sec. 14, T10N, R13E (Bell and Gilmore, 1936: 324-325); Cornish Site, near Louisville (Hill and Cooper, 1937: 293); Gates Site, 2 mi. W Fort Crook, Bellevue (Bell and Gilmore, 1936: 324-325); North Plattsmouth Site, N of Plattsmouth and S of Platte River (Hill and Cooper, 1937: 317); Sheldon Site, Sec. 21, T10N, R13E (Bell and Gilmore, 1936: 324-325); Woolsey Site, Sec. 5, T10N, R14E (Bell and Gilmore, 1936: 324-325). Cedar Co.: Schulte Site, 25CD1, St. Helena (Cooper, 1936: 76); Radke Site, 25CD2, 3 mi. S Wynot (Cooper, 1936: 76; Hill, 1988: 100-123); Wiseman Site, 25CD3, 4.5 mi. NE Wynot (Cooper, 1936: 76; Hill, 1988: 100-123); Ferber Site, 25CD10, near Bow Valley Mill (Ludwickson et al., 1981: 22-23); 25CD36, East Bow Creek, 4.5 mi. S, 1.25 mi. E Wynot (Watson et al., 2004). Chase Co.: Lovitt Site, 25CH1, 12 mi. W Wauneta (Hill and Metcalf, 1942: 204; Gunnerson, 1960: 212-216); Ough Site, 1 mi. E Wauneta (Hill and Kivett, 1940: 232). Cherry Co.: Barrow Pit Site, 25CE231, Fort Niobrara Wildlife Refuge (Osborn, 1979: 51, 111); 25CE309, 3 mi. S, 0.4 mi. W Sparks (Pepperl and Falk, 1983: B84); 25CE312, 3.5 mi. S, 1 mi. W Sparks (Pepperl and Falk, 1983: B87). Cheyenne Co.: Thurston Site, 25CN11, Lodgepole Creek, 6 mi. E Potter (Jensen, 1973: 167); Rock Shelter Site, 25CN23, 4 mi. E Potter (Jensen, 1973: 201); Cave Site, 25CN24, 4 mi. E Potter (Jensen, 1973: 202); 25CN26, Point of Rocks, 4 mi. E Potter (Jensen, 1973: 171); Rock Shelter Site, 25CN27, 4 mi. E Potter (Jensen, 1973: 203); Fred Carlson Site, 25CN29, Lodgepole Creek, 3 mi. E Potter (Jensen, 1973: 176); 25CN55, southeastern Sidney (Carlson, 1999: 105-108); 25CN56, Lodgepole Creek, 1 mi. S Potter (Steinacher and Carlson, 1984: 28). Colfax Co.: Schuyler Site, 25CX1, along Shell Creek, ≈ 3 mi. NW Schuyler (Dunlevy, 1936: 199; Bozell et al., 1982: 28; Graham et al., 1987: 286). Custer Co.: "Doc" McKenzie Site, 25CU92, Spring Creek, 4.6 mi. S, 0.25 mi. E Comstock (Holen and Watson, 1999: 94, 99); 25CU101, Middle Loup River valley, 4.75 mi. S, 0.75 mi. E Comstock (Holen and Watson, 1999: 59). Dakota Co.: Bobier Site, 25DK1, 4.5 mi. SE Homer (Frantz, 1963: 82-83; Hill, 1988: 100-123); 25DK3, 6 mi. W Homer (Price, 1956: 102); Big Village Site, 25DK5, 1 mi. NE Homer (Steinacher and Carlson, 1984: 34-35; O'Shea and Ludwickson, 1992: 332); Nelson Site, 25DK7, 1 mi. N Homer (Frantz, 1963: 82-83; Hill, 1988: 100-123); Hancock Site, 25DK14, 1 mi. SE Homer (Frantz, 1963: 82-83). Dawes Co.: Slaughterhouse Creek Site, 25DW17, western edge of Crawford (Bozell and Ludwickson, 1988: 45);

25DW19, 0.75 mi. W Crawford (Bozell and Ludwickson, 1988: 72); 25DW21, 0.6 mi. S, 2 mi. W Crawford (Bozell and Ludwickson, 1988: 74); 25DW59, 0.70 mi. W Crawford (Bozell and Ludwickson, 1988: 86); Big Cottonwood #3, 25DW94, Sec. 9, T33N, R52W (Meston, 1976: 25); 25DW102, Sec. 31, T33N, R52W (Meston, 1976: 31-32); FN-2, 2.5 mi. S, 3 mi. E Crawford (Bozell and Pepperl, 1987: 69). Dawson Co.: 25DS7, Sec. 5, T11N, R19W (Henning, 1977: 114; Ludwickson, 1978b: 35); 25DS19, Sec. 3, T8N, R20W, right of way for I-80 near Platte River (Frantz, 1965: 20); 25DS21, Platte River valley, 4.75 mi. S, 4 mi. E Lexington (Garrett, 1965: 75-76; Roper, 1994: 62); 25DS102, Sec. 36, T10N, R19W (Henning, 1977: 101); 25DS107, 1.25 mi. W Sumner (Ludwickson, 1978b: 36); 25DS108, Sec. 1, T11N, R 19W (Ludwickson, 1978b: 36); 25DS109, 6 mi. SW Overton (Ludwickson, 1978b: 36); 25DS110, Sec. 2, T11N, R19W (Ludwickson, 1978b: 32); 25DS117, Sec. 16, T10N, R19W (Ludwickson, 1978b: 36). Dixon Co.: Peterson Site, 25DX1, near Ponca (Hill, 1988: 100-123); 25DX2, 3 mi. S, 2 mi. W Ponca (Price, 1956: 123); Annie's Site, 25DX30, in vicinity of 2 mi. N, 3 mi. E Newcastle (Giessen, 1988: 133-156; Manz and Blakeslee, 1988: 159-160). Douglas Co.: Champe Site, 25DO1, margin of uplands overlooking bottom lands through which flow the Elkhorn and Platte rivers (Hill and Cooper, 1936b: 267); Parker Site, 25DO2, on the bluffs north of Ponca Creek and west of the Missouri River, in extreme northeastern Douglas Co. (Ewing, 2000: 2); unnamed site, approximately 15th and I Streets, Omaha (Gilder, 1909: 78). Deuel Co.: Neumann Site, 25DU3, Lodgepole Creek, 3.5 mi. SE Chappell (Carlson, 1973: 104). Franklin Co.: Lost Creek Site, 25FR6, Lost Creek, 2.25 mi. S, 1.5 mi. E Bloomington (Strong, 1935: 94-100); Holdrege 3, head of Lost Creek, 6 mi. S Bloomington (Wedel, 1934: 157, 161, 201); Holdrege 4, Rebecca Creek, 6 mi. SW Bloomington (Wedel, 1934: 161, 201); Red Cloud 2, 4 mi. SE Franklin (Wedel, 1934: 149, 201); Red Cloud 3, 3.5 mi. S, 1.5 mi. E Franklin (Wedel, 1934: 152, 201). Frontier Co.: Owens Site, 25FT3, on Medicine Creek (Mick, 1983: 172; Graham et al., 1987: 285); Dick Site, 25FT9, between the confluence of Brush Creek and Medicine Creek, 2 mi. W Curtis (Gunnerson 1960: 222); 25FT13, 14, 16, 17, vicinity of Medicine Creek Dam (Mick, 1983: 175, 182. 185, 191, 210; Graham et al., 1987: 285-286; Kivett and Metcalf, 1997: 213; Nepstad-Thornberry et al., 2002: 199); 25FT18, 19, 20, 28, 30, near point where Lime Creek originally entered Medicine Creek (Mick, 1983: 202; Graham et al., 1987: 286; Kivett and Metcalf, 1997: 212-213); 25FT22, ≈ 4 km upstream on Medicine Creek from Medicine Creek Dam (Turnmire, 1996: 189-191; Kivett and Metcalf, 1997: 213; Nepstad-Thornberry et al., 2002: 199); Spring Creek Site, 25FT31, Red Willow Creek, adjacent to Red Willow Dam (Kivett, 1961: 10-11; Grange, 1980: 166, 170; Widga, 2004: 29); 25FT32, Red Willow Creek, 0.75 mi. upstream from Red Willow Dam (Grange, 1980: 166, 170; Mick, 1983: 169); Mowry Bluff Site, 25FT35, 6 mi. NW Cambridge (Falk, 1969a: 42, 1969b: 44; Mick, 1983: 205; Graham et al., 1987: 286; Nepstad-Thornberry et al., 2002: 199); 25FT36, ≈ 4 km upstream on Medicine Creek from Medicine Creek Dam (Mick, 1983: 207; Kivett and Metcalf, 1997: 213); 25FT39, ≈ 5 km upstream on Medicine Creek from Medicine Creek Dam (Mick, 1983: 210; Graham et al., 1987: 286; Kivett and Metcalf, 1997: 213); Lime Creek Site, 25FT41, north side of Lime Creek valley, 1 mi. from original junction of Lime and Medicine creeks (Davis, 1962: 23; Graham et al., 1987: 283; Wedel, 1986: 69); Allen Site, 25FT50, ≈ 3 mi. upstream from Medicine Creek Dam (Holder and Wike, 1949: 261; Bamforth, 2002: 65; Hudson, 2007: 195); 25FT54, Red Willow Creek, 5 mi. upstream from Red Willow Dam (Grange, 1980: 170; Mick, 1983: 169); 25FT58, Red Willow Creek, ≈4.5 mi. upstream from Red Willow Dam (Grange, 1980: 171); 25FT77, Red Willow Creek, ≈1 mi. upstream from Red Willow Dam (Grange, 1980: 171); 25FT80, Red Willow Creek, 4.5 mi. upstream from Red Willow Dam (Grange, 1980: 171; Mick, 1983: 169); Medicine Creek 4, 2 mi. S Stockville (Wedel, 1934: 166, 168, 201). Garden Co.: Ash Hollow Cave Site, 25GD2, 3 mi. SE Lewellen (Champe, 1946: 43); O. V. Clary site, 25GD50, Ash Hollow Draw, 4.25 mi. S, 1 mi. W Lewellen (Hill et. al., 2008: 125-126); 25GD54, Blue Creek, 2 mi. N, 9 mi. E Oshkosh (Bozell et al., 2008: 59); 25GD84, Blue Creek, 2.5 mi. N, 9.75 mi. E Oshkosh (Bozell et al., 2008: 82); Clary Ranch Site, 25GD106, Ash Hollow Draw, 4.25 mi. S Lewellen (Hill, 2005: 254-255; Hill et. al., 2008: 117); Barn Bluff, 6.5 mi. SE Oshkosh (Renaud, 1933: 23-24, 48); unnamed site, North Platte River, 4.5 mi. W Oshkosh (Renaud, 1933: 19, 48). Garfield Co.: 25GF13, 1 mi. S, 2.1 mi. W Burwell (Bozell and Pepperl, 1982: 10, 12); 25GF14, 1 mi. S, 1.5 mi. W Burwell (Bozell and Pepperl, 1982: 15). Gosper Co.: Wallace Site, 25GO2, Plum Creek, 5.5 mi. N, 9.5 mi. E Elwood (Winfrey, 1991: 83); Mousel Site (Turkey Creek I), 25GO5, 16 mi. S, 5 mi. W Elwood (Philips, 1957a: 3); Stillman Site, 25GO35, about 15 mi. S, 8 mi. W Elwood (Philips, 1960b: 2); Helms Site, 25GO39, 14 mi. S, 9 mi. W Elwood (Philips, 1956a: 1); Greisfeller Site, 25GO50/51, 13 mi. S, 10 mi. W Elwood (Philips, 1960c: 1-2); Reuben Lucas Site, 25GO52, 12 mi. S, 10 mi. W Elwood (Philips, 1958: 2); Gross Site, 25GO53, 12 mi. S, 10 mi. W Elwood (Philips, 1956b: 2); Lowell Dawson Site, 25GO87, Sec. 1, T5N, R22W (Philips, 1959: 2); Dawson Site, 25GO88, Sec. 12, T5N, R22W (Philips, 1960d: 2-3, 5). Greeley Co.: Schudel Site, 25GY12, North Loup River valley, vicinity Scotia (Roper, 1998: 27). Hall Co.: 25HL3-25HL4, Sec. 29, T9N, R12W, right of way for I-80 near Platte River (Frantz, 1965: 40); Hulme Site, 25HL28, 16 mi. W Grand Island (Bozell, 1991: 234). Harlan Co.: Sindt Point, 25HN16, 3 mi. S, 2.75 mi. E Alma (Adair and Brown, 1987: 115, 573); 25HN36, 2 mi. S Alma (Adair and Brown, 1987: 154, 586); White Cat Village, 25HN37, Prairie Dog Creek, 6 mi. SE Alma (Champe, 1949: 289;

Gunnerson, 1960: 177; Adair and Brown, 1987: 168, 588); Green Plum Site, 25HN39, 5.5 mi. S, 0.8 mi. E Alma (Rusco, 1960: 40); Blue Stone Site, 25HN45, Prairie Dog Creek, 7 mi. SE Alma (Rusco, 1960: 40); 25HN54, 3 mi. S, 2.5 mi. E Alma (Adair and Brown, 1987: 237, 592); 25HN57, 4.5 mi. S, 3 mi. E Alma (Adair and Brown, 1987: 243, 593); Tipover Cove Site, 25HN178, 1.25 mi. S, 3 mi. E Alma (Holen et al. 1996: 46); 25HN180, 2 mi. S, 3.5 mi. E Alma (Holen et al., 1996: 15); 25HN182, 4.2 mi. S, 5.9 mi. E Alma (Holen et al., 1996: 15). Hitchcock Co.: Massacre Canyon Site, 25HK13, ≈ 3 mi. E Trenton (Kivett, 1951: 35; Kivett, 1952: 21). Holt Co.: 25HT1, 17 mi. N, 3 mi. W O'Neill (Price, 1956: 81); 25HT2, Redbird Creek, 22 mi. NE O'Neill (Wood, 1956: 64; 1965: 107-108, 112); Redbird Site, 25HT3, 0.2 mi. N Redbird (Wood, 1956: 40-42, 45;1965: 107-108, 112). Hooker Co.: Lowe Site, 25HO07, 1 1/2 mi. N, 1 mi. E Mullen (Gunnerson, 1960:186); Humphrey Site, 25HO21, Middle Loup River, about 5 mi. E Mullen (Kivett, 1947: 28; Gunnerson, 1960: 198-204); Kelso Site, 25HO23, Middle Loup River, about 5 mi. E Mullen (Kivett, 1952: 39-40). Howard Co.: Palmer Site, 4 mi. N, 1 mi. W Palmer (Wedel, 1936: 83); Stabaco Site, 25HW16, Loup River, near Cushing (Goedert, 1995: 170); Schmidt Site, 25HW301, along North Loup River, near Elba (Mick, 1983: 157; Graham et al., 1987: 284). Keith Co.: 25KH17, Sec. 11, T13N, R39W (Gant, 1966: 3); 25KH73, Sec. 13, T13N, R37W (Holen and Watson, 2000: 67); 25KH83, Brule Canyon, near Brule (Holen and Watson, 2000: 73). Keya Paha Co.: 25KP243, East Middle Creek, 2.25 mi. S, 1.25 mi. E Norden (Pepperl and Falk, 1983: B91); 25KP261, 0.5 mi. S, 2.5 mi. W Norden (Pepperl and Falk, 1983: B94). Knox Co.: Ponca Fort, 25KX1, 1 mi. E Verdel (Wood, 1993: 40, 42); Minarik Site, 25KX2, 9, Ponca Creek, 7 mi. NW Verdel (Wood, 1956: 113-115, 119-120, 137, 140; 1965: 107-109, 112); Wirth Site, 25KX4, Niobrara River, 14 mi. upstream from confluence of Niobrara and Missouri rivers (Wood, 1956: 79; 1965: 107-109, 112); 25KX15, about 4 mi. N, 19.5 mi. Niobrara (Blakeslee and O'Shea, 1983: 177-179); 25KX30, Santee (Carlson, 1976: 4); 25KX52, east bank Weigand Creek, about 4 mi. N, 24 mi. E Niobrara (Blakeslee and O'Shea, 1983: 179-181); 25KX53, near entrance Weigand Creek Recreation Area, about 4 mi. N, 24 mi. E Niobrara (Blakeslee and O'Shea, 1983: 182); 25KX56, mouth of Devils Nest Creek, about 3.75 mi. N, 15 mi. E Niobrara (Blakeslee and O'Shea, 1983: 189); 25KX57, mouth Weigand Creek, about 4.5 mi. N, 24 mi. E Niobrara (Blakeslee and O'Shea, 1983: 196, 197-198); 25KX59, near Bloomfield Recreation Area, about 4.5 mi. N, 21 mi. E Niobrara (Blakeslee and O'Shea, 1983: 204); 25KX64, about 3 mi. N, 7.5 mi. E Niobrara (Blakeslee and O'Shea, 1983: 209); 25KX67, about 2.5 mi. N, 6.5 mi. Niobrara (Blakeslee and O'Shea, 1983: 214); 25KX68, about 2 mi. N, 6.25 mi. E Niobrara (Blakeslee and O'Shea, 1983: 216); 25KX69, about 2 mi. N, 6.25 mi. E Niobrara (Blakeslee and O'Shea, 1983: 217-218); 25KX71, about 3 mi. E Niobrara (Blakeslee and O'Shea, 1983: 221-222); 25KX200, about 4.5 mi. N, 23 mi. E Niobrara; (Frantz, 1965: 50; Blakeslee and O'Shea, 1983: 233-234); 25KX203, Deepwater Recreation Area, about 4.5 mi. N, 25 mi. E Niobrara (Blakeslee and O'Shea, 1983: 239-242); Tramp Deep Site, 25KX204, about 5 mi. E Niobrara; (Howard and Gant, 1966: 18, 23, ; Blakeslee and O'Shea, 1983: 242-245). Lancaster Co.: Schrader Site, Salt Creek, near Roca, about 9 mi. S Lincoln (Hill and Cooper, 1936a: 244, 246). Lincoln Co.: Keith Ranch Site, 25LN34, abandon Keith Ranch, south of South Platte River (Carlson, 1973: 32). Loup Co.: Blowout Site, 25LP7, 0.75 mi. upstream on Dry Creek from its confluence with Calamus River (Falk and Pepperl, 1981: A1-51); Royal Goodenow Site, 25LP8, confluence of Dry Creek and Calamus River, 8 mi. NW Burwell (Falk and Pepperl, 1981: A1-80-81; Roper, 1989: 313-316); Gaylord Wallace Site, 25LP14, 0.75 mi. upstream on Gracie Creek from its confluence with Calamus River (Falk and Pepperl, 1981: A1 123); Russel L. Bald Site, 25LP17, 1.5 mi. upstream on Dry Creek from its confluence with the Calamus River (Falk and Pepperl, 1981: A1 159). Madison Co.: 25MD3, near Union Creek, 1.75 mi. E Madison (Steinacher and Carlson, 1984: 53-55). Merrick Co.: Tahaksu Site, 25MK15, 4.8 mi. N, 1.2 mi. W Palmer (Watson, 1996: 135). Morrill Co.: 25MO14, Lower Dugout Creek Drainage, Sec. 9, T19N, R47W (Bozell, 2008: 157-158); 25MO37, Lower Dugout Creek Drainage, Sec. 3, T19N, R47W (Bozell et al., 2008: 160); 25MO169, Greenwood Creek, ≈ 5 mi. SE Bridgeport (Bozell et al., 2008: 139-140); Greenwood Site, old Greenwood Stage Station, Keenan Ranch, ≈ 9 mi. S, 8 mi. E Redington (Renaud, 1933: 14); Site 2, 1 mi. S head of Cedar Creek to the south of Broadwater (Bell and Cape, 1936: 382); Site 45, ≈ 9 mi. S, 8 mi. E Redington (Bell and Cape, 1936: 382). Nance Co.: Burkett Site, 4 mi. SW Genoa (Blackman, 1907: 339, 341; 1924: 3; Dunlevy, 1936: 200); 25NC8, 10, 12, 13, vicinity of Fullerton (Ludwickson, 1978a: 98; Mick, 1983: 160, 163, 166; Graham et al., 1987: 285); Palmer Locality [in part], 25NC29, 5.5 mi. S, 19 mi. W Fullerton (Meadow and Peterson, 2001: 157); unnamed site, 3 mi. E Genoa (Wedel, 1936: 83-84). Nemaha Co.: Leahy Site, 25NH1, ≈ 3 mi. W Peru (Hill and Kivett, 1940: 199); Majors Site, 25NH2, ≈ 3 mi. NW Peru (Hill and Cooper, 1937: 324); Osborne Site, 25NH5, Missouri River, N of Brownsville (Hill and Cooper, 1937: 338); McKissick Site, 3 mi. NW Peru (Hill and Cooper, 1937: 349; Hill and Kivett, 1940: 199). Nuckolls Co.: 25NO2, Sec. 3, T4N, R7W, terrace about 1/4 mi. Blue River (Frantz, 1965: 60). Otoe Co.: 25OT24, Sec. 26, T8N, R14E (Kaldahl, 1993: 106-112). Pawnee Co.: Table Rock Site, 25PW1, Sec. 33, T3N, R12E (Bell and Gilmore, 1936: 324-325). Platte Co.: Bakenhus Site, 25PT2, Loseke Creek about 50 m below forks of Elk and Deer creeks (Lamb, 1939a: 35-39; Hill and Kivett, 1940: 212); Bakenhus ossuary, 25PT4, Sec. 15, T19N, R1E (Lamb, 1939b: 11); Hill-Rupp Site, 25PT13, 1.5 mi. N Monroe (Metcalf, 1941a: 25, 30, 34). Red Willow Co.:

25RW22, 3 mi. N, 6.25 mi. W Indianola (Grange 1980: 166, 172; Mick, 1983: 169); 25RW25, 3.25 mi. N, 6.25 mi. W Indianola (Grange, 1980: 172); Doyle Site, 25RW28, 9.5 mi. N, 0.75 mi. W McCook (Grange, 1980: 172); Gillen Pits, 25RW101, 4.5 mi. W McCook (Corner, 1977: 79); Davidson Pits, 25RW102, 1 mi. W Bartley (Corner, 1977: 79, 86); Redwillow 1, 6 mi. NW McCook (Wedel, 1934: 171-172, 201). Richardson Co.: Leary Village Site, 25RH1, 4 mi. SE Rulo (Hill and Wedel, 1936: 63-64). Sarpy Co.: Childs Point District, 25SY0, ≈ 1.5 mi. NW Bellevue (Gilder, 1909: 71, 73); Fremont 1, 25SY1, margin of uplands overlooking bottom lands through which flow the Elkhorn and Platte rivers (Hill and Cooper, 1936c: 287, 292); Fontenelle Forest Site, 25SY3, near Bellevue (Hill and Cooper, 1937: 309); Gates Site, 25SY5, probably near modern Capehart Road and 36th Street in southern Bellevue (Strong, 1935:153); Lucien Fontenelle's Post, 25SY26, Bellevue (Bozell et al., 1990: 30); Patterson Site, 25SY31, ≈ 7 mi. S Gretna (Bozell and Ludwickson, 1999: 84); unnamed site, near Meadow (Brooking, 1921: 38-39). Scotts Bluff Co.: Signal Butte Site, 25SF1, Signal Butte, 11.5 mi. S Morrill (Strong, 1935: 235-236); Scottsbluff Bison Quarry, 25SF2, 16 mi. W, 3 mi. S Scottsbluff (Schultz and Eiseley, 1935: 306-307; Hill, 2008: 90; Knudson, 2013: 290); 25SF11, Scotts Bluff National Monument, summit Scotts Bluff, 0.45 mi. N, 0.3 mi. E Visitor Center (Bozell, 1994b: 50, 63); 25SF39, Scotts Bluff National Monument, 0.2 mi. S, 0.2 mi. W Visitor Center (Bozell, 1994b: 50-51, 63); 25SF50, Scotts Bluff National Monument, 0.9 mi. N, 0.4 mi. E Visitor Center (Cannon, 1991: 59-60); Scott's Spring, 25SF51, Scotts Bluff National Monument, 0.2 mi. N, 0.4 mi. E Visitor Center (Bozell, 1994b: 51, 63); 25SF53, 3.25 mi. S, 0.75 mi. E Lyman (Koch and Miller, 1996: 36); 25SF65, Scotts Bluff National Monument, 0.7 mi. N, 0.1 mi. E Visitor Center (Bozell, 1994b: 51, 63); 25SF76, Scotts Bluff National Monument, 0.7 mi. N, 0.1 mi. W Visitor Center (Bozell, 1994b: 52, 63). Sherman Co.: Bill Packer Site, 25SM9, along Davis Creek in extreme northeastern corner of county (Graham et al., 1987: 285; Bozell and Rogers, 1989: 28); Sherman Feeder Canal, 25SM203, Sec. 5, T16N, R15W (Cumming and Stephenson, 1953: 6); Sweetwater Site, 0.5 mi. NW Sweetwater [in Buffalo Co.] (Champe, 1936: 267-268). Sioux Co.: 25SX26, Agate Fossil Beds National Monument, 0.4 mi. N, 0.6 mi. E Visitor Center (Bozell, 1993: 60); Cedar Canyon #1-#3, #7, 25SX103, Sec. 16, T33N, R53W (Meston, 1976: 21-23); Hudson-Meng Site, 25SX115, 23 mi. NW Crawford (Agenbroad, 1978: 61); Whitehead West #2, 25SX116, Sec. 13, T33N, R54 (Meston, 1976: 18-19); Sand Creek 3-4, 25SX141, Sec. 28, T33N, R 53W (Meston, 1976: 37); Sand Creek 3-5, 25SX142, Sec. 28, T33N, R53W (Meston, 1976: 38); Milo Saxton, 25SX146, Sec. 28, T33N, R53W (Meston, 1976: 35-36); 25SX153, Agate Fossil Beds National Monument, 0.5 mi. W Visitor Center (Bozell, 1993: 60-61); 25SX157, Agate Fossil Beds National Monument, 19.25 mi. S, 5.75 mi. E Harrison (Bozell, 1993: 61; Bozell, 1994a: 48, 62); 25SX163, Agate Fossil Beds National Monument, 19.25 mi. S, 7.25 mi. E Harrison (Kay, 1975: 43; Bozell, 1993: 64); 25SX268, Agate Fossil Beds National Monument (Galindo, 1997: 122); 25SX472, Agate Fossil Beds National Monument, 0.65 mi. E Agate (Bozell, 1993: 65); 25SX475, Agate Fossil Beds National Monument, 1.5 mi. E Agate (Bozell, 1993: 66); 25SX476, Agate Fossil Beds National Monument, 1.7 mi. E Agate (Bozell, 1993: 67); 25SX486, Agate Fossil Beds National Monument, 19.25 mi. S, 5.25 mi. E Harrison (Bozell, 1994a: 50, 63). Stanton Co.: Stanton Site, 25ST1, 1 mi. E Stanton (Gunnerson, n.d.). Thayer Co.: Rose Creek, Sec. 29, T1N, R1W (Lamb, 1932: 171). Thurston Co.: Gerald Rose Site, 25TS10, vicinity of Winnebago (Gant, 1966: 6; Carlson, 1973: 53-54); Omaha Creek Site, Sec. 27, T26N, R8E (Gant, 1966: 6). Valley Co.: Mira Creek Site, 25VY1, 6 mi. W, 0.75 mi. N North Loup (Hill and Kivett, 1940: 183). Washington Co.: Site W1, on Douglas-Washington Co. line in extreme southeastern corner of county (Cooper, 1939: 111). Webster Co.: Hill Site, 25WT1, 2 mi. S, 7 mi. E Red Cloud (Wedel, 1936: 62; Metcalf, 1941b: 16, 34); Shipman Site, 25WT7, south side of Republican River, between Red Cloud and Guide Rock (Mick, 1983: 154; Graham et al., 1987: 286); Superior 2, 3 mi. W Guide Rock (Wedel, 1934: 143, 201); Red Cloud 1, Red Cloud (Wedel, 1934:148, 201).

Remarks.—Bison were abundant and statewide in distribution in times prior to European settlement (Jones, 1964). Not surprisingly then bison are known from nearly every Holocene archeological and paleontological site in Nebraska. Clearly this species was abundant in Nebraska during the Holocene and probably statewide in its distribution, although it may have been less abundant in the eastern-most parts of the state. Bison were recovered from 276 Holocene sites in 60 of Nebraska's 93 counties. The exact distribution of bison in Nebraska based on archeological sites is made more difficult because native peoples butchered and transported many of their kills over a considerable distance before consumption. The presence of skulls or nearly complete skeletons may be the best indication that the bison was taken in the local area.

It is believed that most if not all of the Holocene bison in Nebraska belonged to *B. b. bison*. However, Agenbroad (1978: 61) believed that older material (about 10,000 years old), such as from his Hudson-Meng Site (25SX115) in Sioux Co. might be assigned to either of the fossil taxa *B. b. occidentalis* or *B. b. antiquus*. There are a number of other Holocene sites in Nebraska that are as old or approach the age of the Hudson-Meng material. The sites with bison remains that may be 9000 years old or older include the following: 25CN56, Cheyenne Co.; 25DW102, Dawes Co.; Lime Creek site, 25FT41, and Allen site, 25FT50, Frontier Co.; O.V. Clary site, 25GD50, and Clary Ranch site, 25GD106, Garden Co.; 25HN180 and 25HN182, Harlan Co; Gillen Pits, 25RW101, and Davidson Pits, 25RW102, Red Willow Co.; and Scottsbluff Bison Quarry, 25SF2, Scotts Bluff Co. Of note is that all of these Holocene sites are from three counties in southwestern Nebraska along or near the Republican River and in four counties (counting Hudson-Meng) in the Panhandle of western Nebraska. In addition, bison provide a complete record throughout the Holocene in Nebraska.

ORDER RODENTIA

***Sciurus niger*—eastern fox squirrel**

Site record (1).—Gosper Co.: Wallace Site, 25GO2, Plum Creek, 5.5 mi. N, 9.5 mi. E Elwood (Winfrey, 1991: 82; Souders, 1994: 53).

Remarks.—The fox squirrel is one of two species of tree squirrels occurring in Nebraska. It is found in the forests of eastern Nebraska, but in the western third of the state the species is restricted to riparian and urban forests. In these situations fox squirrels are a forest edge species using both the trees and surrounding open areas. It is thought that fox squirrels were not present in the western parts of Nebraska until settlement resulted in the development of riparian and urban parks (Jones, 1964; Genoways et al., 2008b). This makes the record from Gosper Co. of particular interest because it is believed to be from between AD 1450 to AD 1675. This material should be carefully re-examined to verify its identification.

***Sciurus* sp.—tree squirrel**

Site records (6).—Cass Co.: Walker Gilmore, 25CC28, 6 mi. SE Murray (Haas, 1983: 247). Colfax Co.: Schuyler Site, 25CX1, along Shell Creek, ≈ 3 mi. NW Schuyler (Bozell et al., 1982: 27; Graham et al., 1987: 286). Douglas Co.: Cabanne's Post, 25DO8, near Dodge Park northeast of Omaha (Bozell et al., 1990: 32). Otoe Co.: 25OT24, Sec. 26, T8N, R14E (Kaldahl, 1993: 106-112). Sarpy Co.: Lucien Fontenelle's Post, 25SY26, Bellevue (Bozell et al., 1990: 29); Patterson Site, 25SY31, ≈ 7 mi. S Gretna (Bozell and Ludwickson, 1999: 83).

Remarks.— There are two members of the tree squirrel genus *Sciurus* in Nebraska—*S. niger* and *S. carolinensis*. The skeletons of these two species are difficult to separate without the critical diagnostic parts. The species are essentially the same size, are closely related, and are best identified using external and dental characters.

As I stated above, *S. niger* is essentially statewide in its modern distribution, whereas *S. carolinensis* is confined to the mature forests bordering the Missouri River south of Omaha. The two species could have been present at four of these sites, with Schuyler site and Patterson sites being along the Platte River and thus outside of the modern range of *S. carolinensis*; however, the Patterson site is not far west of the junction of the Platte and Missouri rivers. These remains could be insightful into the early distribution of tree squirrels in Nebraska if they can be identified to species. The remains from the Walker Gilmore site are the oldest, being as much as 1300 years old.

Urocitellus elegans—Wyoming ground squirrel

Site record (1).—Harlan Co.: 25HN180, 2 mi. S, 3.5 mi. E Alma (Holen et al., 1996: 15).

Remarks.—This record of a ground squirrel was reported under the scientific name *Spermophilus richardsonii* (Holen et al., 1996); therefore, the current name usage may be confusing. For many years *S. richardsonii* and *S. elegans* were believed to be members of the same species (Zegers, 1984); however, the work of Nadler et al. (1971) and subsequent researchers has demonstrated that *S. richardsonii* and *S. elegans* should be considered distinct species. The geographic range of the former species is to the north and northwest of Nebraska, whereas the latter species is found primarily to the west of Nebraska with two historic records from the southern Panhandle (Jones, 1964).

In a 2009, a major study of the systematics of ground squirrels of the genus *Spermophilus* was published (Helgren et al., 2009). Ground squirrels formerly considered to be members of a single genus—*Spermophilus*—were shown to represent eight separate genera and the genus *Spermophilus* was restricted to Old World ground squirrels. There are now at least four genera of Nebraska ground squirrels as I have used in this account and those following.

This is an interesting record, being approximately 360 km [225 mi.] southeast of the nearest modern record of the species at Bridgeport in western Nebraska and about 480 km [300 mi.] east of populations in central Colorado. The record is from an early Holocene site with the age of the deposits being 9700 to 10,500 years ago. The presence of the Wyoming ground squirrel at this site may indicate a cooler and drier environment at the time.

Poliocitellus franklinii—Franklin's ground squirrel

Site records (9).—Brown Co.: McIntosh Site, 25BW15, Enders Lake, ≈12 mi. S, 12.5 mi. W Ainsworth (Koch, 1995: 43; 2004: 116). Butler Co.: White Site, 25BU20 (Ore, 1981: 11); Palmer Johnson Site, 25BU37, east of Ulysses (Souders, 1994: 53; Bozell and Koch, n.d.). Colfax Co.: Schuyler Site, 25CX1, along Shell Creek, ≈ 3 mi. NW Schuyler (Bozell et al., 1982: 27; Graham et al., 1987: 286). Douglas Co.: Cabanne's Post, 25DO8, near Dodge Park northeast of Omaha (Bozell et al., 1990: 32). Hall Co.: Hulme Site, 25HL28, 16 mi. W Grand Island (Bozell, 1991: 233; Souders, 1994: 53). Howard Co.: Schmidt Site, 25HW301, along North Loup River, near Elba (Satorius-Fox, 1982: 8; Mick, 1983: 155; Graham et al., 1987: 284). Sarpy Co.: Patterson Site, 25SY31, ≈ 7 mi. S Gretna (Bozell and Ludwickson, 1999: 83). Sherman Co.: Bill Packer Site, 25SM9, along Davis Creek in extreme northeastern corner of county (Graham et al., 1987: 284; Bozell and Rogers, 1989: 27; Souders, 1994: 53).

Remarks.—The modern geographic distribution of Franklin's ground squirrel in Nebraska includes the eastern half of the state and then westward along river systems as far west as a place near Lisco, Morrill Co., along the North Platte River (Jones, 1964; Genoways et al., 2008b). This species has long been associated with the tall grass prairie, but recent work by Huebschman (2007) has shown that these ground squirrels prefer habitats including both grass and woody vegetation that would best be called parkland. These squirrels hibernate over winter in a deep burrow system.

All nine Holocene sites for Franklin's ground squirrel fall within the modern geographic range of the species in Nebraska. The site in Brown Co., which is at the very edge of the modern geographic range lies between the Niobrara and North Loup rivers. Based on the ages of these nine sites, it appears that Franklin's ground squirrel has occupied it current range in Nebraska for at least 2000 years.

Ictidomys tridecemlineatus—thirteen-lined ground squirrel

Site records (14).—Boone Co.: Beaver Creek Site, 25BO23, 4.5 mi. S Petersburg (Koch and Nelson, 2002: 86). Brown Co.: McIntosh Site, 25BW15, Enders Lake, ≈12 mi. S, 12.5 mi. W Ainsworth (Koch, 1995: 43; 2004: 116). Butler Co.: White Site, 25BU20 (Ore, 1981: 11). Frontier Co.: Owens Site, 25FT3, on Medicine Creek (Mick, 1983: 171; Graham et al., 1987: 284); 25FT14, 17, vicinity of Medicine Creek Dam (Kivett and Metcalf, 1997: 213); 25FT22, Sec. 11, T5N, R26W (Turnmire, 1996: 189-191); Mowry Bluff Site, 25FT35, 6 mi. NW Cambridge (Falk, 1969b: 47; Mick, 1983: 204; Graham et al., 1987: 286). Hall Co.: Hulme Site, 25HL28, 16 mi. W Grand Island (Bozell, 1991: 233; Souders, 1994: 53). Harlan Co.: 25HN36, 2 mi. S Alma (Adair and Brown, 1987: 153, 582). Howard Co.: Schmidt Site, 25HW301, along North Loup River, near Elba (Satorius-Fox, 1982: 8; Mick, 1983: 155; Graham et al., 1987: 284). Keya Paha Co.: 25KP311, 5.25 mi. S, 1 mi. E Norden (Pepperl and Falk, 1983: B94). Sarpy Co.: Patterson Site, 25SY31, ≈ 7 mi. S Gretna (Bozell and Ludwickson, 1999: 83). Sherman Co.: Bill Packer Site, 25SM9, along Davis Creek in extreme northeastern corner of county (Graham et al., 1987: 284; Bozell and Rogers, 1989: 27; Souders, 1994: 53).

Remarks.—The thirteen-lined ground squirrel is statewide in distribution, being a common grasslands inhabitant. It prefers short-grass situations where it is able to see above the grass while standing on its hind legs (Jones, 1964; Genoways et al., 2008b). These squirrels hibernate over winter in a deep burrow system.

Holocene sites for this ground squirrel are confined to the eastern half of Nebraska, with sites in Brown Co. and Frontier Co. being the western-most. The Holocene sites cover only the time period of the last 2100 years. The White site in Butler Co. probably contains the oldest material from the state. These data add very little to our current understanding of this species in Nebraska.

Ictidomys, Poliocitellus, Urocitellus, Xerospermophilus sp.—ground squirrel

Site records (12).—Burt Co.: Logan Creek Site A, B, C, 25BT3, Logan Creek, SW Oakland (Graham et al., 1987: 283; Widga, 2006: 67). Cass Co.: Walker Gilmore, 25CC28, 6 mi. SE Murray (Haas, 1983: 252). Frontier Co.: 25FT17, vicinity of Medicine Creek Dam (Mick, 1983: 189; Graham et al., 1987: 286); 25FT22, Sec. 11, T5N, R26W (Turnmire, 1996: 189-191). Harlan Co.: 25HN65, 1.25 mi. W Alma (Moore, 1988: 145). Nance Co.: Palmer Locality [in part], 25NC29, 5.5 mi. S, 19 mi. W Fullerton (Meadow and Peterson, 2001: 156). Otoe Co.: 25OT24, Sec. 26, T8N, R14E (Kaldahl, 1993: 106-112). Sherman Co.: Bill Packer Site, 25SM9, along Davis Creek in extreme northeastern corner of county (Graham et al., 1987: 284; Bozell and Rogers, 1989: 27). Sioux Co.: 25SX163, Agate Fossil Beds National Monument, 19.25 mi. S, 7.25 mi. E Harrison (Bozell, 1993: 63). Webster Co.: Shipman Site, 25WT7, south side of Republican River, between Red Cloud and Guide Rock (Mick, 1983: 153; Graham et al., 1987: 286).

Remarks.—With four species of long-tailed ground squirrels in Nebraska (the three species discussed above, plus the spotted ground squirrel, *Xerospermophilus spilosoma*), it is extremely difficult to impossible to identify fragmentary remains to the species level. Even though these four species are representatives of four distinct genera incomplete skeletons are difficult to separate because the best distinguishing characters involve genetic materials and external characteristics. Size, at times, is helpful, but without data for age and secondary sexual variations, it is of limited value. It is, therefore, not surprising that a significant amount of Holocene ground squirrel material remains unidentified.

Cynomys ludovicianus—black-tailed prairie dog

Site records (39).—Boone Co.: Beaver Creek Site, 25BO23, 4.5 mi. S Petersburg (Koch and Nelson, 2002: 86). Brown Co.: McIntosh Site, 25BW15, Enders Lake, ≈12 mi. S, 12.5 mi. W Ainsworth (Koch, 1995: 43; 2004: 116). Burt Co.: Logan Creek Site B, C, D, 25BT3, Logan Creek, SW Oakland (Snyder and Bozell, 1983: 16; Graham et al., 1987: 283; Widga, 2006: 67-68). Cherry Co.: 25CE309, 3 mi. S, 0.4 mi. W Sparks (Pepperl and Falk, 1983: B85). Colfax Co.: Schuyler Site, 25CX1, along Shell Creek, ≈ 3 mi. NW Schuyler (Bozell et al., 1982: 27; Graham et al., 1987: 286). Franklin Co.: Lost Creek Site, 25FR6, Lost Creek, 2.25 mi. S, 1.5 mi. E Bloomington (Strong, 1935: 100). Frontier Co.: Owens Site, 25FT3, on Medicine Creek (Mick, 1983: 171; Graham et al., 1987: 284); 25FT13, 16, 17, 70, vicinity of Medicine Creek Dam (Mick, 1983: 173, 184, 190, 209; Graham et al., 1987: 284; Kivett and Metcalf, 1997: 213); 25FT18, 20, 30, near point where Lime Creek originally entered Medicine Creek (Mick, 1983: 201; Graham et al., 1987: 286; Kivett and Metcalf, 1997: 212-213); 25FT22, SE ¼, Sec. 11, T5N, R26W (Turnmire, 1996: 189-191); Spring Creek Site, 25FT31, Red Willow Creek, adjacent to Red Willow Dam (Grange, 1980: 166; Widga, 2004: 29); Mowry Bluff Site, 25FT35, 6 mi. NW Cambridge (Falk, 1969b: 47; Mick, 1983: 204; Graham et al., 1987: 286); 25FT39, ≈ 5 km upstream on Medicine Creek from Medicine Creek Dam (Mick, 1983: 209; Graham et al., 1987; Kivett and Metcalf, 1997: 213); Lime Creek Site, 25FT41, north side of Lime Creek valley, 1 mi. from original junction of Lime and Medicine creeks (Davis, 1962: 23; Wedel, 1986: 69); Allen Site, 25FT50, ≈ 3 mi. upstream from Medicine Creek Dam (Holder and Wike, 1949: 261; Bamforth, 2002: 65; Hudson, 2007: 195); 25FT77, Red Willow Creek, ≈1 mi. upstream from Red Willow Dam (Grange, 1980: 171); 25FT80, Red Willow Creek, 4.5 mi. upstream from Red Willow Dam (Grange, 1980: 171; Mick, 1983: 168). Garden Co.: Ash Hollow Cave Site, 25GD2, 3 mi. SE Lewellen (Champe, 1946: 43-44). Gosper Co.: Wallace Site, 25GO2, Plum Creek, 5.5 mi. N, 9.5 mi. E Elwood (Winfrey, 1991: 82; Souders, 1994: 53). Hall Co.: Hulme Site, 25HL28, 16 mi. W Grand Island (Bozell, 1991: 233; Souders, 1994: 53). Harlan Co.: 25HN36, 2 mi. S Alma (Adair and Brown, 1987: 153, 582). Hooker Co.: Kelso Site, 25HO23, Middle Loup River, about 5 mi. E Mullen (Kivett, 1952: 39-40). Howard Co.: Schmidt Site, 25HW301, along North Loup River, near Elba (Satorius-Fox, 1982: 8; Mick, 1983: 155; Graham et al., 1987: 284). Merrick Co.: Tahaksu Site, 25MK15, 4.8 mi. N, 1.2 mi. W Palmer (Watson, 1996: 135). Morrill Co.: Greenwood Site, old Greenwood Stage Station, Keenan Ranch, ≈ 9 mi. S, 8 mi. E Redington (Renaud, 1933: 14). Nance Co.: Palmer Locality [in part], 25NC29, 5.5 mi. S, 19 mi. W Fullerton (Meadow and Peterson, 2001: 156). Otoe Co.: 25OT24, Sec. 26, T8N, R14E (Kaldahl, 1993: 106-112). Red Willow Co.: 25RW22, 3 mi. N, 6.25 mi. W Indianola (Grange 1980: 172; Mick, 1983: 168); Doyle Site, 25RW28, 9.5 mi. N, 0.75 mi. W McCook (Grange, 1980: 172). Sherman Co.: Bill Packer Site, 25SM9, along Davis Creek in extreme northeastern corner of county (Graham et al., 1987: 284; Bozell and Rogers, 1989: 27; Souders, 1994: 53). Sioux Co.: 25SX268, Agate Fossil Beds National Monument (Galindo, 1997: 122). Valley Co.: Mira Creek Site, 25VY1, 6 mi. W, 0.75 mi. N North Loup (Hill and Kivett, 1940: 183).

Remarks.—Prior to European settlement the black-tailed prairie dog was a common resident of the short-grass and mid-grass prairies of Nebraska (Jones, 1964; Genoways et al., 2008b). The species did not occupy the eastern 10 to 15% of the state where tall-grass prairie and trees predominated. Since that time because of habitat loss and rigorous pest control programs, the black-tailed prairie dog populations in Nebraska have been reduced to no more than 2% of their original size.

The Holocene records for the black-tailed prairie dog in Nebraska coincide closely with the historic distribution of the species. There are records throughout the state except for the northeast and extreme southeast. Of the 14 counties that border the Missouri River, there are Holocene records from only two of the counties. One of the sites (25BT3) is in western Burt County along Logan Creek, southwest of Oakland. This is one of the older Holocene sites in the state with this material being about 7000 years old. Although this site may be east of historic records, it is only a short distance from the eastern-most modern site reported by Jones (1964) at Clarke Creek, near Fontanelle, Washington Co. Both of these sites are west of the bluffs along the Missouri River and are not directly on the river. The other site, located at 25OT24, Sec. 26, T8N, R14E in Otoe Co., is approximately 112 km [70 mi.] east of the nearest modern locality in the area listed by Jones (1964) in Saline Co. This Holocene material from Otoe Co. is much younger than that from the other eastern Holocene sites in Burt Co., being only 650 to 1100 years old. The data seem to indicate that black-tailed prairie dogs have occupied a relatively stable geographic range in Nebraska throughout the Holocene and into modern times until intensive agriculture displaced them. The eastern boundary of the geographic range of the species has not been a fixed line, but probably fluctuated from time to time with changing environmental conditions that broke up or pushed to the east the areas of forest and tall-grass prairie.

These records for black-tailed prairie dogs basically cover the whole time period of the Holocene. The oldest of the records come from the Allen site (25FT50) in Frontier Co. where the material is 9500 to 13,000 years old and this is followed closely by material from the Lime Creek site also in Frontier Co., which has been dated to 8000 to 9500 years old. The Logan Creek site in Burt Co., as mentioned above, is about 7000 years old. The Spring Creek site (25FT31) adjacent to the Red Willow Dam in Frontier Co. dates from 5500 to 5850 years ago, whereas all remaining sites are 2100 years old or younger.

Watson (1996) reported some material from a place in Merrick Co.—Tahaksu Site, 25MK15, 4.8 mi. N, 1.2 mi. W Palmer—as *Cynomys leucurus*, the white-tailed prairie dog. Modern populations of this species live in the high, dry mountain valleys in southern Montana, western Wyoming, northwestern Colorado, and northeastern Utah. It seems unlikely to me that this species was a resident of east-central Nebraska about 700 to 1000 years ago. I have not included an account for it in this summary of Holocene records.

***Marmota monax*—woodchuck**

Site records (2).—Frontier Co.: 25FT13, vicinity of Medicine Creek Dam (Mick, 1983: 173; Graham et al., 1987: 284); 25FT16, vicinity of Medicine Creek Dam (Mick, 1983: 184; Graham et al., 1987: 284; Kivett and Metcalf, 1997: 213).

Remarks.—The woodchuck is a common inhabitant of the woodlands and forest edges in the eastern United States. Nebraska lies at the western edge of the distribution of the species in the central United States. Roehrs and Genoways (2004) documented the expansion of the geographic range of the woodchuck in Nebraska during the twentieth century. They demonstrated that the species had expanded its range from a limited area in the eastern part of the state to now occur in at least the eastern half of the state. The species has been expanding its geographic ranges about 4 km per year over this period. This westward dispersal of the woodchuck has followed the expansion of the riparian forests along the rivers and creeks in Nebraska. These riparian forests provide vast areas of forest edges, which is a preferred habitat of the woodchuck. The woodchuck's expansion westward across Nebraska has been slower than some other eastern species, such as the white-footed mouse, *Peromyscus leucopus*.

The two Holocene records of woodchucks in Nebraska are both from Frontier Co. in the southwestern part of the state. These sites are beyond the modern geographic range of woodchuck even with the expansions that have occurred in the 20th century. The nearest documented modern site for the woodchuck in southern Nebraska is 1 mile south and 0.5 mile west of Franklin, Franklin Co. about 112 km [70 mi.] to the southeast of the Medicine Creek Dam (Roehrs and Genoways, 2004). The modern record is on the Republican River and the Holocene records are along Medicine Creek, which is a major tributary of the Republican River. Both Holocene sites date to 1000 to 1100 years ago. These facts would seem to indicate that about 1000 years ago there was a riparian forest well enough developed along Medicine Creek to provide a forest edge habitat. The most likely route for this population to reach Frontier Co. would have been for individuals to move along the Republican River from Kansas where it connected with the Kansas River and then the Missouri River. However, the expansion of the geographic range of the woodchuck in Kansas has mirrored what has occurred in Nebraska in the 20th century (Choate and Haner, 1992; Wilson and Choate, 1996; Kaufman et al., 2016). The nearest record in Kansas of a woodchuck in the Republican River valley was taken in Jewell Co. in 1985 (Choate and Reed, 1986) so modern records in Kansas have not approached Frontier Co. any nearer than Nebraska records.

Another potential explanation for this record is that it does not pertain to the woodchuck, but rather to its close relative the yellow-bellied marmot, *Marmota flaviventris*, which occurs in the Rocky Mountains in Colorado. The nearest modern record to Frontier Co., Nebraska, is from near Fort Collins about 415 km [256 mi.] to the west. Although this explanation may seem like a remote possibility, Stewart (1987: 301, 316) reported *Marmota* cf. *flaviventris* from a late Pleistocene deposit at the North Cove site (25HN164) Harlan County Lake, Harlan County. This place is not far geographically from the sites in Frontier Co., but they are separated by more than 10,000 years in time. Two other Holocene species, *Urocitellus elegans* and *Thomomys talpoides*, have populations in this general area in Colorado.

A lack of Holocene records is another point of interest for woodchucks in Nebraska. It is the third largest rodent in Nebraska after the American beaver and porcupine, with some individuals weighing as much as 5 kg. Woodchucks are a game animal that is still consumed in parts of the United States so one would expect that woodchucks would have been included in the native diet when the opportunity arose. In my mind, this argues for woodchucks being confined to the extreme eastern parts of Nebraska along the Missouri River prior to European settlement.

Castor canadensis—American beaver

Site records (61).—Buffalo Co.: Trampee's Bend Site, 25BF171, 1.5 mi. SE Amherst (Bleed, 1973: 119-121). Burt Co.: Logan Creek Site B, 25BT3, Logan Creek, SW Oakland (Snyder and Bozell, 1983: 17; Graham et al., 1987: 283; Widga, 2006: 67). Butler Co.: Linwood Site, 25BU1, Skull Creek, 1 mi. SW Linwood (Carlson, 1973: 79); Barcal Site, 25BU4, near Abie (O'Shea and Ludwickson, 1992: 338). Cass Co.: Theodore Davis Site, 25CC17, 1.5 mi. E Weeping Water (Gradwohl, 1969: 143); Walker Gilmore, 25CC28, 6 mi. SE Murray (Haas, 1983: 251; Graham et al., 1987: 284). Cedar Co.: Ferber Site, 25CD10, near Bow Valley Mill (Ludwickson et al., 1981: 22-23). Chase Co.: Lovitt Site, 25CH1, 12 mi. W Wauneta (Hill and Metcalf, 1942: 204; Gunnerson, 1960: 212-216). Cherry Co.: 25CE309, 3 mi. S, 0.4 mi. W Sparks (Pepperl and Falk, 1983: B85). Colfax Co.: Schuyler Site, 25CX1, along Shell Creek, ≈ 3 mi. NW Schuyler (Dunlevy, 1936: 199; Bozell et al., 1982: 27; Graham et al., 1987: 286). Dakota Co.: Big Village Site, 25DK5, 1 mi. NE Homer (O'Shea and Ludwickson, 1992: 331); 25DK10, 1 mi. N Homer (O'Shea and Ludwickson, 1992: 234). Douglas Co.: Parker Site, 25DO2, on the bluffs north of Ponca Creek and west of the Missouri River, in extreme northeastern Douglas Co. (Ewing, 2000: 4); Cabanne's Post, 25DO8, near Dodge Park northeast of Omaha (Bozell et al., 1990: 32). Franklin Co.: Lost Creek Site, 25FR6, Lost Creek, 2.25 mi. S, 1.5 mi. E Bloomington (Strong, 1935: 100). Frontier Co.: Owens Site, 25FT3, on Medicine Creek (Mick, 1983: 171; Graham et al., 1987: 284); 25FT13, 14, 16, 17, 70, vicinity of Medicine Creek Dam (Mick, 1983: 174, 182, 184, 190, 209; Graham et al., 1987: 284, 286; Kivett and Metcalf, 1997: 213; Nepstad-Thornberry et al., 2002: 199); 25FT18, 20, 30, near point where Lime Creek originally entered Medicine Creek (Mick, 1983: 201; Graham et al., 1987: 286; Kivett and Metcalf, 1997: 212-213); 25FT22, 36, ≈ 4 km upstream on Medicine Creek from Medicine Creek Dam (Mick, 1983: 206; Turnmire, 1996: 189-191; Kivett and Metcalf, 1997: 213; Nepstad-Thornberry et al., 2002: 199); Spring Creek Site, 25FT31, Red Willow Creek, adjacent to Red Willow Dam (Grange, 1980: 170; Widga, 2004: 29); 25FT32, Red Willow Creek, 0.75 mi. upstream from Red Willow Dam (Grange, 1980: 166; Mick, 1983: 168); Mowry Bluff Site, 25FT35, 6 mi. NW Cambridge (Falk, 1969b: 47; Mick, 1983: 205; Graham et al., 1987: 286; Nepstad-Thornberry et al., 2002: 199); 25FT39, ≈ 5 km upstream on Medicine Creek from Medicine Creek Dam (Mick, 1983: 209; Graham et al., 1987: 286; Kivett and Metcalf, 1997: 213); Lime Creek Site, 25FT41, north side of Lime Creek valley, 1 mi. from original junction of Lime and Medicine creeks (Davis, 1962: 23; Wedel, 1986: 69; Graham et al., 1987: 283); Allen Site, 25FT50, ≈ 3 mi. upstream from Medicine Creek Dam (Holder and Wike, 1949: 261; Hudson, 2007: 195); 25FT80, Red Willow Creek, 4.5 mi. upstream from

Red Willow Dam (Grange, 1980: 171; Mick, 1983: 168). Garden Co.: Ash Hollow Cave Site, 25GD2, 3 mi. SE Lewellen (Champe, 1946: 43-44). Gosper Co.: Wallace Site, 25GO2, Plum Creek, 5.5 mi. N, 9.5 mi. E Elwood (Winfrey, 1991: 82); Dawson Site, 25GO88, Sec. 12, T5N, R22W (Philips, 1960d: 3). Hall Co.: Hulme Site, 25HL28, 16 mi. W Grand Island (Bozell, 1991: 233). Harlan Co.: 25HN36, 2 mi. S Alma (Adair and Brown, 1987: 153, 583); White Cat Village, 25HN37, Prairie Dog Creek, 6 mi. SE Alma (Champe, 1949: 289; Gunnerson, 1960: 177). Hooker Co.: Humphrey Site, 25HO21, Middle Loup River, about 5 mi. E Mullen (Gunnerson, 1960: 204); Kelso Site, 25HO23, Middle Loup River, about 5 mi. E Mullen (Kivett, 1952: 39-40). Howard Co.: Stabaco Site, 25HW16, Loup River, near Cushing (Goedert, 1995: 170); Schmidt Site, 25HW301, along North Loup River, near Elba (Satorius-Fox, 1982: 8; Mick, 1983: 155; Graham et al., 1987: 284). Loup Co.: Royal Goodenow Site, 25LP8, confluence of Dry Creek and Calamus River, 13 km NW Burwell [in Garfield Co.] (Falk and Pepperl, 1981: A1-80). Morrill Co.: Greenwood Site, old Greenwood Stage Station, Keenan Ranch, ≈ 9 mi. S, 8 mi. E Redington (Renaud, 1933: 14). Nance Co.: 25NC8, 10, 13, vicinity of Fullerton (Mick, 1983: 159, 162, 165; Graham et al., 1987: 284); Palmer Locality [in part], 25NC29, 5.5 mi. S, 19 mi. W Fullerton (Meadow and Peterson, 2001: 157). Platte Co.: Bakenhus ossuary, 25PT4, Sec. 15, T19N, R1E (Lamb, 1939b: 11). Red Willow Co.: 25RW25, 3.25 mi. N, 6.25 mi. W Indianola (Grange, 1980: 172); Doyle Site, 25RW28, 9.5 mi. N, 0.75 mi. W McCook (Grange, 1980: 167, 173). Richardson Co.: Leary Village Site, 25RH1, 4 mi. SE Rulo (Hill and Wedel, 1936: 63); Rulo Southeast, 25RH69, 0.4 km S Rulo (Nelson, 2006: 165); Rulo Southeast, 25RH70, 0.6 km S Rulo (Nelson, 2006: 170, 173). Sarpy Co.: Childs Point District, 25SY0, ≈ 1.5 mi. NW Bellevue (Gilder, 1909: 72-73); Lucien Fontenelle's Post, 25SY26, Bellevue (Bozell et al., 1990: 29); Patterson Site, 25SY31, ≈ 7 mi. S Gretna (Bozell and Ludwickson, 1999: 83); unnamed site, near north side of Bellevue (Gilder, 1909: 78). Stanton Co.: Stanton Site, 25ST1, 1 mi. E Stanton (Gunnerson, n.d.). Valley Co.: Mira Creek Site, 25VY1, 6 mi. W, 0.75 mi. N North Loup (Hill and Kivett, 1940: 183). Webster Co.: Shipman Site, 25WT7, south side of Republican River, between Red Cloud and Guide Rock (Mick, 1983: 153; Graham et al., 1987: 286).

Remarks.—This semi-aquatic rodent occurred statewide prior to European settlement. They created dams on creeks and rivers to form ponds where their lodges could be located. Those beavers living on larger lakes and rivers made burrows in the banks to hold their nests. The demand for beaver pelts for making of felt top hats and other clothing during the 1800s resulted in a nearly complete depletion of the species in Nebraska and in most other areas of it geographic range. The situation was critical by 1907 when laws were passed giving the beaver complete protection in Nebraska. Populations in Nebraska did rebound so that after 1952 there has been a trapping season each year. Populations of beavers are not large in Nebraska but they may be expected in any aquatic habitat (Jones, 1964; Genoways et al., 2008b).

Beaver skeletal material is among the commonest rodent remains in Holocene sites in Nebraska. Beavers are the largest rodent in the state and were used by Native Americans both for food and clothing. Their pelage is dense, soft, and water repellent. Davis (1962) in describing the oldest faunal zone at the early Holocene site of Lime Creek in Frontier Co. stated: "Beaver and pronghorn bones make up most of the material, a circumstance which makes this zone unusual among Early Lithic sites in the Plains." Younger deposits at this site are dominated by bison bones as other Plains sites. Davis (1962) theorized that the Lime Creek valley at this time (9500 years ago) was swampy, with "a near-by beaver pond" which made beavers the most available small game.

The 61 Nebraska Holocene sites for beavers occur in 28 counties generally covering all regions of the state. The area with the least coverage is the northern tier of counties that is in the Niobrara River drainage. There is only a single record here in the vicinity of Sparks in Cherry Co. The oldest Holocene remains of beaver in Nebraska are those from the Allen site in Frontier Co., where the bones may be as old as 13,000 years. The remainder of the Holocene is covered by other records reported here with the youngest material being a little less than 200 years old at Cabanne's Post, Douglas Co., and Lucien Fontenelle's Post, Sarpy Co.

Perognathus sp.—pocket mice

Site records (6).—Frontier Co.: 25FT17, vicinity of Medicine Creek Dam (Kivett and Metcalf, 1997: 213); Lime Creek Site, 25FT41, north side of Lime Creek valley, 1 mi. from original junction of Lime and Medicine creeks (Graham et al., 1987: 283). Howard Co.: Schmidt Site, 25HW301, along North Loup River, near Elba (Satorius-Fox, 1982: 8; Graham et al., 1987: 284). Loup Co.: Gracie Creek Site, 25LP16, 0.5 mi. upstream on Gracie Creek from its confluence with the Calamus River (Falk and Pepperl, 1981: A1 151). Richardson Co.: Rulo Southeast, 25RH70, 0.6 km S Rulo (Nelson, 2006: 170, 173).

Remarks.—Pocket mice were traditionally classified in a single genus, *Perognathus*, until the work of Hafner and Hafner (1983) who divided them into two genera *Perognathus* (silky pocket mice) and *Chaetodipus* (hispid pocket mice). There are three species of silky pocket mice known to live in Nebraska, with two being about the same size and the third smaller and quite rare. The plains pocket mouse, *Perognathus flavescens*, occurs almost statewide, but is not found in the southeast (Reed and Choate, 1986). It is about the same size as the second species, both being about 130 to 140 millimeters in total length. The second medium-size pocket mouse is *P. fasciatus*, olive-backed pocket mouse, which occurs in the Panhandle and a small area of sand hills north of Niobrara River. The third species is the smallest (total length, 110 millimeters) and rarest silky pocket mouse, *P. flavus*, which is found in the Panhandle and western Sand Hills.

Based only on distribution, these records would best be assigned to *Perognathus flavescens*, but characteristics of the bones will be needed for definite identification. The site in Richardson Co. in extreme southeastern Nebraska is beyond the geographic range of the three species of *Perognathus*, but is closest to *P. flavescens*. This species occurs in areas of prairie and is most abundant on the sandy soils of the Sand Hills (Jones, 1964; Genoways et al., 2008b).

The Lime Creek site in Frontier Co. is among the older Holocene sites in Nebraska, with material dating to 8000 to 9500 years ago. At the opposite end of the Holocene time scale are remains from the Gracie Creek site in Loup Co., where deposits dated from AD 1850 to AD 1910.

Chaetodipus hispidus—hispid pocket mouse

Site records (6).—Boone Co.: Beaver Creek Site, 25BO23, 4.5 mi. S Petersburg (Koch and Nelson, 2002: 86). Frontier Co.: Owens Site, 25FT3, on Medicine Creek (Mick, 1983: 171; Graham et al., 1987: 284); 25FT22, Sec. 11, T5N, R26W (Turnmire, 1996: 189-191). Gosper Co.: Wallace Site, 25GO2, Plum Creek, 5.5 mi. N, 9.5 mi. E Elwood (Winfrey, 1991: 82; Souders, 1994: 53). Howard Co.: Schmidt Site, 25HW301, along North Loup River, near Elba (Satorius-Fox, 1982: 8; Mick, 1983: 155; Graham et al., 1987: 284). Sherman Co.: Bill Packer Site, 25SM9, along Davis Creek in extreme northeastern corner of county (Graham et al., 1987: 284; Bozell and Rogers, 1989: 27; Souders, 1994: 53).

Remarks.—The hispid pocket mouse at present is statewide in its distribution, preferring areas of prairie, pastureland, and sandy areas (Jones, 1964; Genoways et al., 2008b). This is the sole representative of the genus *Chaetodipus* in Nebraska. It can be distinguished from the pocket mice of the *Perognathus* in the state by it larger size (total length, 220 mm), details of the cranial anatomy, and harsher dorsal pelage.

With only six Holocene records, we learn very little about the distribution of hispid pocket mice in this era. All of this material is from relatively young sites, with the oldest material from about 1500 years ago from the Wallace site in Gosper Co.

Chaetodipus sp.—pocket mouse

Site record (1).—Nance Co.: Palmer Locality [in part], 25NC29, 5.5 mi. S, 19 mi. W Fullerton (Meadow and Peterson, 2001: 157).

Remarks.—These bones must belong to *C. hispidus* because it is the only member of the genus in Nebraska. All other members of the genus are confined to the southwestern U. S. and into Mexico (Hall, 1981). These remains are from between AD 1100-AD 1400.

Dipodomys ordii—Ord's kangaroo rat

Site records (4).—Boone Co.: Beaver Creek Site, 25BO23, 4.5 mi. S Petersburg (Koch and Nelson, 2002: 86). Frontier Co.: 25FT17, vicinity of Medicine Creek Dam (Mick, 1983: 190; Graham et al., 1987: 286; Kivett and Metcalf, 1997: 213); 25FT22, Sec. 11, T5N, R26W (Turnmire, 1996: 189-191); Mowry Bluff Site, 25FT35, 6 mi. NW Cambridge (Falk, 1969b: 47; Mick, 1983: 204; Graham et al., 1987: 286).

Remarks.—Ord's kangaroo rats occupy the western two-thirds of Nebraska, with the eastern-most modern record at 6 miles south of Monroe, Platte Co. This places the eastern boundary of the geographic range of the species at approximately the 98th meridian. Kangaroo rats prefer sandy soils so are particularly abundant in the Sand Hills. They eat the seeds of grasses, forbs, and similar plants.

With remains at only four sites, there is not much that we can learn about the Holocene distribution of Ord's kangaroo rat; however, it is interesting that the site in Boone Co. is almost exactly on the modern eastern boundary of the species. This would indicate that Ord's kangaroo rats have been this far east in Nebraska for at least the last 1000 years. The sites in Frontier Co. date from the same time period as the Boone Co. site. The Frontier Co. sites fall well within the modern distribution of *D. ordii*.

Geomys sp.—pocket gopher

Site records (73).—Boone Co.: Beaver Creek Site, 25BO23, 4.5 mi. S Petersburg (Koch and Nelson, 2002: 86). Brown Co.: McIntosh Site, 25BW15, Enders Lake, ≈12 mi. S, 12.5 mi. W Ainsworth (Koch, 1995: 43; 2004: 117); 25BW249, 17 mi. N, 6.5 mi. W Johnstown (Pepperl and Falk, 1983: B81); 25BW250, 16.5 mi. N, 6 mi. W Johnstown (Pepperl and Falk, 1983: B81). Burt Co.: Logan Creek Site A, B, C, 25BT3, Logan Creek, SW Oakland (Snyder and Bozell, 1983: 16; Graham et al., 1987: 283; Widga, 2006: 67). Butler Co.: Linwood Site, 25BU1, Skull Creek, 1 mi. SW Linwood (Carlson, 1973: 79); White Site, 25BU20 (Ore, 1981: 11); Palmer Johnson Site, 25BU37, east of Ulysses (Souders, 1994: 53; Bozell and Koch, n.d.). Cass Co.: Theodore Davis Site, 25CC17, 1.5 mi. E Weeping Water (Gradwohl, 1969: 143); Walker Gilmore, 25CC28, 6 mi. SE Murray (Haas, 1983: 251; Graham et al., 1987: 284); Little Pawnee Creek Site, 25CC228, 0.8 mi. N, 3 mi. W South Bend (Ludwickson, 1994:116). Cedar Co.: Ferber Site, 25CD10, near Bow Valley Mill (Ludwickson et al., 1981: 22-23). Cherry Co.: 25CE309, 3 mi. S, 0.4 mi. W Sparks (Pepperl and Falk, 1983: B85); 25CE312, 3.5 mi. S, 1 mi. W Sparks (Pepperl and Falk, 1983: B87). Dakota Co.: Big Village Site, 25DK5, 1 mi. NE Homer (O'Shea and Ludwickson, 1992: 331). Dawes Co.: Slaughterhouse Creek Site, 25DW17, western edge of Crawford (Bozell and Ludwickson, 1988: 45); 25DW19, 0.75 mi. W Crawford (Bozell and Ludwickson, 1988: 72); 25DW21, 0.6 mi S, 2 mi. W Crawford (Bozell and Ludwickson, 1988: 74); 25DW59, 0.70 mi. W Crawford (Bozell and Ludwickson, 1988: 86). Dixon Co.: Annie's Site, 25DX30, in vicinity of 2 mi. N, 3 mi. E Newcastle (Manz and Blakeslee, 1988: 159-160). Douglas Co.: Cabanne's Post, 25DO8, near Dodge Park northeast of Omaha (Bozell et al., 1990: 32). Frontier Co.: Owens Site, 25FT3, on Medicine Creek (Mick, 1983: 171; Graham et al., 1987: 284); 25FT13, 14, 16, 17, 70, vicinity of Medicine Creek Dam (Mick, 1983: 174, 182, 184, 190, 209; Graham et al., 1987: 284, 286; Kivett and Metcalf, 1997: 56-58, 213); 25FT22, Sec. 11, T5N, R26W (Turnmire, 1996: 189-191); Spring Creek Site, 25FT31, Red Willow Creek, adjacent to Red Willow Dam (Grange, 1980: 166; Widga, 2004: 29); 25FT32, Red Willow Creek, 0.75 mi. upstream from Red Willow Dam (Grange, 1980: 170; Mick, 1983: 168); Mowry Bluff Site, 25FT35, 6 mi. NW Cambridge (Falk, 1969b: 47; Mick, 1983: 204; Graham et al., 1987: 286); 25FT39, ≈ 5 km upstream on Medicine Creek from Medicine Creek Dam (Mick, 1983: 209; Graham et al., 1987: 286); Allen Site, 25FT50, ≈ 3 mi. upstream from Medicine Creek Dam (Hudson, 2007: 195); 25FT54, Red Willow Creek, 5 mi. upstream from Red Willow Dam (Grange, 1980: 171; Mick, 1983: 168); 25FT58, Red Willow Creek, ≈ 4.5 mi. upstream from Red Willow Dam (Grange, 1980: 171); 25FT80, Red Willow Creek, 4.5 mi. upstream from Red Willow Dam (Grange, 1980: 171; Mick, 1983: 168). Garden Co.: Ash Hollow Cave Site, 25GD2, 3 mi. SE Lewellen (Champe, 1946: 43-44). Gosper Co.: Wallace Site, 25GO2, Plum Creek, 5.5 mi. N, 9.5 mi. E Elwood (Winfrey, 1991: 82). Hall Co.: Hulme Site, 25HL28, 16 mi. W Grand Island (Bozell, 1991: 233). Harlan Co.: 25HN36, 2 mi. S Alma (Adair and Brown, 1987: 153, 582); 25HN40, 0.8 mi. S, 2 mi. E Alma (Adair and Brown, 1987: 194, 590); Rural School District No. 9, 25HN50, 0.8 mi. S, 1.75 mi. E Alma (Adair and Brown, 1985: 211, 592); 25HN65, 1.25 mi. W Alma (Moore, 1988: 145); 25HN125, 0.75 mi. S, 2.1 mi. E Alma (Adair and Brown, 1987: 276, 594). Hooker Co.: Kelso Site, 25HO23, Middle Loup River, about 5 mi. E Mullen (Kivett, 1952: 39-40). Howard Co.: Schmidt Site, 25HW301, along North Loup River, near Elba (Satorius-Fox, 1982: 8; Mick, 1983: 155; Graham et al., 1987: 284). Loup Co.: Royal Goodenow Site, 25LP8, confluence of Dry Creek and Calamus River, 13 km NW Burwell (Falk and Pepperl, 1981: A1-80-81; Roper, 1989: 313-316); Gracie Creek Site, 25LP16, 0.5 mi. upstream on Gracie Creek from its confluence with the Calamus River (Falk and Pepperl, 1981: A1 151). Merrick Co.: Tahaksu Site, 25MK15, 4.8 mi. N, 1.2 mi. W Palmer (Watson, 1996: 135). Nance Co.: 25NC8, 10, vicinity of Fullerton (Mick, 1983: 159, 162; Graham et al., 1987: 284); Palmer Locality [in part], 25NC29, 5.5 mi. S, 19 mi. W Fullerton (Meadow and Peterson, 2001: 157). Red Willow Co.: Doyle Site, 25RW28, 9.5 mi. N, 0.75 mi. W McCook (Grange, 1980: 172) . Richardson Co.: Leary Village Site, 25RH1, 4 mi. SE Rulo (Hill and Wedel, 1936: 63); Rulo Southeast, 25RH69, 0.4 km S Rulo (Nelson, 2006: 165); Rulo Southeast, 25RH70, 0.6 km S Rulo (Nelson, 2006: 170, 173). Sarpy Co.: Childs Point District, 25SY0, ≈ 1.5 mi. NW Bellevue (Gilder, 1909: 72-73); Lucien Fontenelle's Post, 25SY26, Bellevue (Bozell et al., 1990: 29); Patterson Site, 25SY31, ≈ 7 mi. S Gretna (Bozell and Ludwickson, 1999: 83). Scotts Bluff Co.: 25SF11, Scotts Bluff National Monument, summit Scotts Bluff, 0.45 mi. N, 0.3 mi. E Visitor Center (Bozell, 1994b: 50, 63); 25SF76, Scotts Bluff National Monument, 0.7 mi. N, 0.1 mi. W Visitor Center (Bozell, 1994b: 52, 63). Sherman Co.: Bill Packer Site, 25SM9, along Davis Creek in extreme northeastern corner of county (Graham et al., 1987: 284; Bozell and Rogers, 1989: 27; Souders, 1994: 53). Sioux Co.: 25SX157, Agate Fossil Beds National Monument, 19.25 mi. S, 5.75 mi. E Harrison (Bozell, 1993: 61; Bozell, 1994a: 48, 62); 25SX163, Agate Fossil Beds National Monument, 19.25 mi. S, 7.25 mi. E Harrison (Bozell, 1993: 63); 25SX475, Agate Fossil Beds National Monument, 1.5 mi. E Agate (Bozell, 1993: 65); 25SX476, Agate Fossil Beds National Monument, 1.7 mi. E Agate (Bozell, 1993: 66); 25SX486, Agate Fossil Beds National Monument, 19.25 mi. S, 5.25 mi. E Harrison (Bozell, 1994a: 50, 63); 25SX487, Agate Fossil Beds National Monument, 0.1 mi. N, 0.15 mi. E Visitor Center (Bozell, 1993: 67). Valley Co.: Mira Creek Site, 25VY1, 6 mi. W, 3/4 mi. N North Loup (Hill and Kivett, 1940: 183). Washington Co.: Fort Atkinson Site, 25WN9, 0.5 mi. E Fort Calhoun (Carlson, 1979: 124). Webster Co.: Shipman Site, 25WT7, south side of Republican River, between Red Cloud and Guide Rock (Mick, 1983: 153; Graham et al., 1987: 286).

Remarks.—Pocket gophers of the genus *Geomys* are statewide in their distribution in Nebraska. These gophers are highly adapted to fossorial living, meaning they are specialized for burrowing in the ground and they pass most of their lives in these burrows. Members of genus *Geomys* prefer softer, deeper soils that contain sand or loam, but are not restricted to them. Jones (1964) recognized a single species of *Geomys* in Nebraska—*G. bursarius*—and many of the Holocene records have been reported under this name. Unfortunately the systematics of *Geomys* in Nebraska currently is in a state of confusion with three and maybe even four species to be recognized. At the present time, these species are distinguished primarily based on molecular genetic characteristics that are not evident in material recovered from archeological and paleontological sites.

The populations of *Geomys* living in the Sand Hills and Panhandle of Nebraska are characterized by small size and pale pelage coloration. The name currently being applied to this population is *G. lutescens*, which has been used as a species and subspecies name during the history of the taxon (Heaney and Timm, 1983, 1985; Sudman et al., 1987, 2006; Genoways et al., 2008a). Sudman et al. (1987) were the first to present data to indicate that the pocket gophers living in southwestern Nebraska may be a separate species. Subsequently, Sudman et al. (2006)

and Genoways et al. (2008a) have presented additional data in support of this concept. The scientific name applied to this species is *Geomys jugossicularis*, which is slightly larger and darker than *G. lutescens*. This leaves the pocket gopher in the eastern 40% of the state to be known by the taxonomic name *Geomys bursarius*. These populations have larger individuals and are darker in coloration than the western populations (Jones, 1964). Even within this population there are differences in morphology and chromosomal number between northern and southern populations that have not been fully explored (Sudman et al., 1987).

A further complication in assessing Holocene records of *Geomys* is that this burrowing species inadvertently may dig into existing archeological sites and leave their remains. This would mean that these gophers were not associated with human activities and are not from the time period of the site. There seems to be a large number of Holocene records of this species as compared with other rodents, which does give some credence to the idea that some of these Holocene "records" are intrusive.

Pocket gophers of all genera are highly territorial, with males and females of the same species coming together only during breeding season. Also they do not tolerate the presence of other species of gophers, consequently the sympatric occurrence of species of pocket gophers is highly unusual and has been the subject of several studies (for example, Reichman and Baker, 1972; Williams and Baker, 1976; Moulton et al., 1979). It is of interest, therefore, to note that at two Holocene sites—25FT54, Red Willow Creek, 5 mi. upstream from Red Willow Dam, Frontier Co. (Grange, 1980: 171; Mick, 1983: 168), and Doyle Site, 25RW28, 9.5 mi. N, 0.75 mi. W McCook, Red Willow Co. (Grange, 1980: 172)—both the plains pocket gopher, *Geomys* sp., and the northern pocket gopher, *Thomomys talpoides*, have been recovered from the same deposits. The former site dates to AD 1310 and the latter to BC 60 to AD 680. There are *Geomys* remains from some of the earliest Holocene sites such as the Allen site in Frontier Co., which may have remains as old as 13,000 years.

***Thomomys talpoides*—northern pocket gopher**

Site records (9).—Frontier Co.: 25FT13, 14, 16, 17 vicinity of Medicine Creek Dam (Kivett and Metcalf, 1997: 213); 25FT20, near point where Lime Creek originally entered Medicine Creek (Kivett and Metcalf, 1997: 213); 25FT22, ≈ 4 km upstream on Medicine Creek from Medicine Creek Dam (Kivett and Metcalf, 1997: 213); 25FT54, Red Willow Creek, 5 mi. upstream from Red Willow Dam (Grange, 1980: 171; Mick, 1983: 168). Red Willow Co.: 25RW22, 3 mi. N, 6.25 mi. W Indianola (Grange 1980: 172; Mick, 1983: 168); Doyle Site, 25RW28, 9.5 mi. N, 0.75 mi. W McCook (Grange, 1980: 172).

Remarks.—The northern pocket gopher is much smaller in size than the plains pocket gopher, but it is still just as fossorial in its habits. Modern populations of the northern pocket gopher occur in limited areas of the northwestern and southwestern Panhandle of Nebraska. In the northwest, these gophers occur in the Pierre soils north of the Pine Ridge and White River and in the southwest, they occur in the upland areas of the Cheyenne Table. The plains pocket gopher also occurs in the area of the Cheyenne Table, but seems to prefer the deeper soils of the valleys (Jones, 1964; Genoways et al. 2008b). There are no Holocene records of *T. talpoides* from the areas of Nebraska where modern populations occur.

The Holocene records of the northern pocket gopher are concentrated in the area of Red Willow and Frontier cos. in the southwestern corner of the state. There are no known modern records of the northern pocket gopher from this part of Nebraska. The nearest modern populations are about 225 km [140 mi.] to the southwest in the vicinity of Seibert and Flagler,

Kit Carson Co., Colorado. These Colorado localities are from the drainages of the upper tributaries of the Republican River and the Nebraska Holocene sites are also from along tributaries of the Republican River. If individuals of *Geomys* were in the area during the Holocene as they are today, *Geomys* would be expected in the deeper river valley soils and *Thomomys* in the drier and shallower upland soils.

The oldest Holocene remains of the northern pocket gopher come from the Doyle site in Red Willow Co., which has been aged between BC 60 and AD 680. The other eight sites with northern pocket gopher remains all cluster in the time period of AD 1000 to AD 1400. These data would seem to indicate that *Thomomys talpoides* entered southwestern Nebraska about 2150 years ago, remained 1400 to 1500 years, and then sometime less than 700 years ago the population disappeared from the state.

Gradwohl (1969) reported *Thomomys bottae* from the Theodore Davis Site (25CC17) near Weeping Water in Cass Co. in eastern Nebraska based on eight legbones. This is an error. It is certainly not *T. bottae*, which is a montane species of the western United States and approaches Nebraska no closer than central Colorado. If these bones are those of a *Thomomys*, it would be *T. talpoides*, but this seems unlikely given that the nearest Holocene site for *T. talpoides* is about 600 km [375 mi.] west in Frontier Co. and the nearest modern record is from Fort Randall in Gregory Co., South Dakota, 506 km [315 mi.] to the northwest. It seems highly unlikely that *Thomomys* overcame this much unfavorable habitat to reach Cass Co., but rather almost certainly these remains are misidentified *Geomys bursarius*, possibly because of age or secondary sexual dimorphism that was not recognized.

***Zapus hudsonius*—meadow jumping mouse**

Site records (3).—Howard Co.: Schmidt Site, 25HW301, along North Loup River, near Elba (Satorius-Fox, 1982: 8; Graham et al., 1987: 284). Richardson Co.: Rulo Southeast, 25RH69, 0.4 km S Rulo (Nelson, 2006: 165); Rulo Southeast, 25RH70, 0.6 km S Rulo (Nelson, 2006: 170, 173).

Remarks.—The meadow jumping mouse occurs widely in Nebraska, being absent from the southwestern Panhandle, southwestern corner of the state, and from the northwestern Panhandle from the Pine Ridge northward. It occurs in riparian habitats, woodlands, and other mesic situations. The species hibernates in winter for as long as six months in some situations (Jones, 1964; Genoways et al., 2008b).

The three Holocene records of *Zapus* come from well within the modern geographic range of the species in Nebraska. The remains from these sites are from the time period of 400 to 900 years ago.

***Microtus ochrogaster*—prairie vole**

Site records (14).—Boone Co.: Beaver Creek Site, 25BO23, 2.4 km N Loretto (Koch, 2002: i, 1-4; Koch and Nelson, 2002: 87). Brown Co.: McIntosh Site, 25BW15, Enders Lake, ≈12 mi. S, 12.5 mi. W Ainsworth (Koch, 1995: 43; 2004: 117). Frontier Co.: Owens Site, 25FT3, on Medicine Creek (Mick, 1983: 171; Graham et al., 1987: 284); 25FT17, vicinity of Medicine Creek Dam (Kivett and Metcalf, 1997: 213); 25FT22, Sec. 11, T5N, R26W (Turnmire, 1996: 189-191); Mowry Bluff Site, 25FT35, 6 mi. NW Cambridge (Falk, 1969b: 48; Mick, 1983: 205; Graham et al., 1987: 286). Hall Co.: Hulme Site, 25HL28, 16 mi. W Grand Island (Bozell, 1991: 233; Souders, 1994: 53). Harlan Co.: 25HN33, 1.75 S mi. Alma (Adair and Brown, 1987: 135, 575); 25HN36, 2 mi. S Alma

(Adair and Brown, 1987: 153, 583); 25HN62, 4.5 mi. S, 2 mi. E Alma (Adair and Brown, 1987: 262, 593); 25HN65, 1.25 mi. W Alma (Moore, 1988: 145). Howard Co.: Schmidt Site, 25HW301, along North Loup River, near Elba (Satorius-Fox, 1982: 8; Mick, 1983: 156; Graham et al., 1987: 284). Red Willow Co.: Doyle Site, 25RW28, 9.5 mi. N, 0.75 mi. W McCook (Grange, 1980: 173). Sarpy Co.: Patterson Site, 25SY31, ≈ 7 mi. S Gretna (Bozell and Ludwickson, 1999: 83).

Remarks.—The modern distribution of the prairie vole is statewide in Nebraska. It occurs primarily in prairie habitats and is tolerant of arid situations so will be found in upland areas (Jones, 1964; Genoways et al., 2008b).

The Holocene records for the prairie vole are from a scattering of sites in the north, east, southwest, and east-central parts of the state, but that leaves large areas of the state such as the Panhandle and Sand Hills unrepresented where the species currently is common. All of the Holocene sites with prairie vole remains date to the last 2100 years, with those from the Doyle site in Red Willow Co. being the oldest.

***Microtus pennsylvanicus*—meadow vole**

Site records (5).—Butler Co.: Palmer Johnson Site, 25BU37, east of Ulysses (Souders, 1994: 53; Bozell and Koch, n.d.). Frontier Co.: 25FT177, vicinity of Medicine Creek Reservoir (Holen and May, 1989: 5). Garden Co.: Ash Hollow Cave Site, 25GD2, 3 mi. SE Lewellen (Champe, 1946: 43-44). Gosper Co.: Wallace Site, 25GO2, Plum Creek, 5.5 mi. N, 9.5 mi. E Elwood (Winfrey, 1991: 82; Souders, 1994: 53). Howard Co.: Schmidt Site, 25HW301, along North Loup River, near Elba (Satorius-Fox, 1982: 8; Mick, 1983: 156; Graham et al., 1987: 284).

Remarks.—The meadow vole essentially occurs statewide in Nebraska, with the possible exception of an area in the northwestern Panhandle (Jones, 1964; Benedict et al., 2000; Genoways et al., 2008b). The meadow vole prefers grassland and woodlands habitat associated with permanent water. Prairie voles and meadow voles may occur together but meadow voles will be found in the wetter areas and prairie voles will be in the drier areas.

All of the Holocene records fall within the geographic range of the modern populations. However, there is an interesting situation in extreme southwestern Nebraska in Dundy Co. where there was believed to be an isolated subspecies of *M. pennsylvanicus* (Jones, 1964). The records reported by Benedict et al. (2000) essentially filled the geographic area between the main population of meadow voles and the isolated one. It was believed at the time that meadow vole populations extended into this area in the Pleistocene/Holocene, but had retreated northward leaving the isolated population. Subsequently with changing environmental conditions the voles had expanded their geographic range in modern time to again contact the isolated population. The bones from the site in Frontier Co. also in southwestern Nebraska date from the late Pleistocene to 17,000 years ago, which supports the idea that meadow voles had extended into this area during this early period. In addition to this early record, the other four sites with meadow vole remains cover the last 2100 years.

***Microtus* sp.—vole**

Site records (19).—Boone Co.: Beaver Creek Site, 25BO23, 2.4 km N Loretto (Koch, 2002: i, 1-4; Koch and Nelson, 2002: 87). Butler Co.: White Site, 25BU20 (Ore, 1981: 11). Cass Co.: Walker Gilmore, 25CC28, 6 mi. SE Murray (Haas, 1983: 251; Graham et al., 1987: 284). Dawes Co.: Slaughterhouse Creek Site, 25DW17, western edge of Crawford (Bozell and Ludwickson, 1988: 45); 25DW59, 0.70 mi. W Crawford (Bozell and Ludwickson, 1988: 86). Douglas Co.: Cabanne's Post, 25DO8, near Dodge Park northeast of Omaha (Bozell et al., 1990: 32). Frontier Co.: 25FT13, 17, vicinity of Medicine Creek Dam (Mick, 1983: 174, 190; Graham et al., 1987:

284, 286); 25FT18, near point where Lime Creek originally entered Medicine Creek (Kivett and Metcalf, 1997: 212); 25FT22, Sec. 11, T5N, R26W (Turnmire, 1996: 189-191); 25FT31, Red Willow Creek, adjacent to Red Willow Dam (Grange, 1980: 166); Lime Creek Site, 25FT41, north side of Lime Creek valley, 1 mi. from original junction of Lime and Medicine creeks (Davis, 1962: 23; Wedel, 1986: 66-71; Graham et al., 1987: 283); Allen Site, 25FT50, ≈ 3 mi. upstream from Medicine Creek Dam (Hudson, 2007: 195). Gosper Co.: Wallace Site, 25GO2, Plum Creek, 5.5 mi. N, 9.5 mi. E Elwood (Winfrey, 1991: 82). Hall Co.: Hulme Site, 25HL28, 16 mi. W Grand Island (Bozell, 1991: 233). Nance Co.: Palmer Locality [in part], 25NC29, 5.5 mi. S, 19 mi. W Fullerton (Meadow and Peterson, 2001: 157). Richardson Co.: Rulo Southeast, 25RH69, 0.4 km S Rulo (Nelson, 2006: 165). Scotts Bluff Co.: 25SF53, 3.25 mi. S, 0.75 mi. E Lyman (Koch and Miller, 1996: 36). Sherman Co.: Bill Packer Site, 25SM9, along Davis Creek in extreme northeastern corner of county (Graham et al., 1987: 284; Bozell and Rogers, 1989: 28).

Remarks.—As just discussed, there are two species in the genus *Microtus* living over much of Nebraska that are similar in size and habits. To make matters even more complicated, there is a third species often classified in this genus—*Microtus* (*Pitymys*) *pinetorum*—that occurs in extreme eastern Nebraska south of Omaha. None of these species are simple to distinguish based on skeletal materials, but are best identified based on details of cranial and dental anatomy (Jones, 1964; Genoways et al., 2008b).

The Holocene sites in Cass, Douglas, and Richardson cos. could contain the remains of any of these three species of *Microtus*. The remaining 16 sites could have bones from *M. pennsylvanicus* or *M. ochrogaster* or both. The ages of the Holocene sites containing *Microtus* sp. remains cover essentially the complete time period of 13,000 to 400 years ago . The oldest material comes from the Allen site (25FT50), which dates from 13,000 to 9500 years ago, and is succeeded by the Lime Creek site (25FT41) material aging to between 9500 and 8000. The remaining time before the first millennium is covered by two sites—25SF53, 8000 to 2000 years ago and 25FT31, 5850 to 5500 years ago. The remaining sites contain material from the last 2100 years.

Ondatra zibethicus—muskrat

Site records (34).—Boone Co.: Beaver Creek Site, 25BO23, 2.4 km N Loretto (Koch, 2002: i, 1-4; Koch and Nelson, 2002: 87). Brown Co.: McIntosh Site, 25BW15, Enders Lake, ≈12 mi. S, 12.5 mi. W Ainsworth (Koch, 1995: 43; 2004: 117). Cass Co.: Walker Gilmore, 25CC28, 6 mi. SE Murray (Haas, 1983: 251). Colfax Co.: Schuyler Site, 25CX1, along Shell Creek, ≈ 3 mi. NW Schuyler (Bozell et al., 1982: 27; Graham et al., 1987: 286). Dakota Co.: Big Village Site, 25DK5, 1 mi. NE Homer (O'Shea and Ludwickson, 1992: 331). Douglas Co.: Parker Site, 25DO2, on the bluffs north of Ponca Creek and west of the Missouri River, in extreme northeastern Douglas Co. (Ewing, 2000: 4); Cabanne's Post, 25DO8, near Dodge Park northeast of Omaha (Bozell et al., 1990: 32). Frontier Co.: 25FT16, 17, 70, vicinity of Medicine Creek Dam (Mick, 1983: 184, 190, 210; Graham et al., 1987: 284, 286; Kivett and Metcalf, 1997: 213); 25FT22, Sec. 11, T5N, R26W (Turnmire, 1996: 189-191); 25FT30, near point where Lime Creek originally entered Medicine Creek (Mick, 1983: 201; Graham et al., 1987: 286); Mowry Bluff Site, 25FT35, 6 mi. NW Cambridge (Falk, 1969b: 48; Mick, 1983: 205; Graham et al., 1987: 286); 25FT36, ≈ 4 km upstream on Medicine Creek from Medicine Creek Dam (Mick, 1983: 206; Kivett and Metcalf, 1997: 213); 25FT39, ≈ 5 km upstream on Medicine Creek from Medicine Creek Dam (Mick, 1983: 210; Graham et al., 1987: 286); 25FT80, Red Willow Creek, 4.5 mi. upstream from Red Willow Dam (Grange, 1980: 171; Mick, 1983: 168). Gosper Co.: Wallace Site, 25GO2, Plum Creek, 5.5 mi. N, 9.5 mi. E Elwood (Winfrey, 1991: 82). Hall Co.: Hulme Site, 25HL28, 16 mi. W Grand Island (Bozell, 1991: 233). Harlan Co.: 25HN36, 2 mi. S Alma (Adair and Brown, 1987: 153, 583); Indian Hill, 25HN42, 3.75 mi. S, 4 mi. E Alma (Adair and Brown, 1987: 207, 591). Hooker Co.: Kelso Site, 25HO23, Middle Loup River, about 5 mi. E Mullen (Kivett, 1952: 39-40). Howard Co.: Schmidt Site, 25HW301, along North Loup River, near Elba (Satorius-Fox, 1982: 8; Graham et al., 1987: 284). Knox Co.: 25KX200, about 4.5 mi. N, 23 mi. E Niobrara; (Blakeslee and O'Shea, 1983: 233-234). Merrick Co.: Tahaksu Site, 25MK15, 4.8 mi. N, 1.2 mi. W Palmer (Watson, 1996: 135). Nance Co.: 25NC8, 10, vicinity of Fullerton (Mick, 1983: 159; Graham et al., 1987: 284); Palmer Locality [in part], 25NC29, 5.5 mi. S, 19 mi. W Fullerton (Meadow and Peterson, 2001: 157). Red Willow Co.: 25RW22, 3 mi. N, 6.25 mi. W Indianola (Grange 1980: 166, 172; Mick, 1983: 168); Gillen Pits, 25RW101, 4.5 mi. W McCook (Corner, 1977: 79-80). Richardson Co.: Rulo Southeast, 25RH69, 0.4 km S Rulo (Nelson, 2006: 165). Sarpy Co.: Lucien Fontenelle's Post, 25SY26, Bellevue (Bozell et al., 1990: 29); Patterson Site, 25SY31, ≈ 7 mi. S Gretna (Bozell and Ludwickson, 1999: 83). Sherman Co.: Bill Packer Site, 25SM9, along Davis Creek in extreme northeastern corner of county (Graham et al., 1987: 284; Bozell and Rogers, 1989: 28). Stanton Co.: Stanton Site, 25ST1, 1 mi. E Stanton (Gunnerson, n.d.). Valley Co.: Mira Creek Site, 25VY1, 6 mi. W, 0.75 mi. N North Loup (Hill and Kivett, 1940: 183).

Remarks.—The muskrat is a semi-aquatic rodent that occurs throughout Nebraska in appropriate habitats (Jones, 1964; Genoways et al., 2008b). These appropriate habitats always include a permanent water source. In lakes, ponds, and similar situations, muskrats will construct a lodge of cattails, reeds, or other aquatic vegetation. Along rivers, streams, or places with only limited aquatic vegetation, muskrats will place their nests in burrows that they have dug into a bank. In any situation in which muskrats place their nests, the entrance burrow will be located below the water surface. The muskrat is the fourth largest rodent in Nebraska, being only smaller than the beaver, porcupine, and woodchuck (Jones, 1964; Genoways et al., 2008b).

Muskrat remains occur in 34 Holocene sites in Nebraska, which is a relatively high number of sites. Muskrats were probably not used as a food source by native people, but pelts of muskrats would have been valuable in the making of clothing because of their waterproof qualities. The sites are grouped geographically in a pattern that is not similar to other species. There are sites in six counties along the Missouri River, eight counties in the east-central part of the state, four counties in the southwest, and two counties in the Sand Hills (Brown and Hooker). This leaves large areas of the state unrepresented and does not reflect the modern distribution of muskrats. The oldest site with remains of the muskrat is Gillen Pits in Red Willow Co. where the material originated from the late Pleistocene to the early Holocene. All other sites with muskrat remains fall in the last 2100 years.

Synaptomys cooperi—southern bog lemming

Site records (2).—Frontier Co.: Allen Site, 25FT50, ≈ 3 mi. upstream from Medicine Creek Dam (Hudson, 2007: 195). Howard Co.: Schmidt Site, 25HW301, along North Loup River, near Elba (Satorius-Fox, 1982: 8; Mick, 1983: 156; Graham et al., 1987: 284).

Remarks.—The southern bog lemming is widely distributed but is nowhere abundant in the eastern half of Nebraska (Jones, 1964;Wilson and Choate, 1997; Genoways et al., 2008b). There is an isolated population of southern bog lemmings in Dundy Co. in extreme southwestern Nebraska that seems to be restricted to the grassy habitats along Rock Creek. As this indicates, the preferred habitat of the southern bog lemming in Nebraska is moist soils with dense ground cover of grass with "an overstory of tall grasses, sunflowers, sumac, wild plum, or the like" (Jones, 1964). More recently, Benedict et al. (2000) reported modern specimens taken from owl pellets found in Red Willow Co. in southwestern Nebraska. This narrowed the gap to the isolated population to just 125 km [78 mi.].

The Holocene record from Howard Co. is from near the center of the modern distribution of the southern bog lemming in Nebraska and dates from the time period of 1000 to 550 years ago. The Holocene record from Frontier Co. is of real interest because it falls between the main body of the bog lemming population in Nebraska and the isolated population in Dundy Co. and it is near the recently discovered populations in Red Willow Co. This Holocene site is about 134 km [83 mi.] ENE of the Rock Creek population. The Allen site is also of interest because it is one of the earliest Holocene sites in Nebraska falling in the range of 9500 to 13,000 years ago.

It would not take much of an environmental change for the southern bog lemming to have extended a continuous population along the Republican River and its tributaries as far west as Rock Creek in Dundy Co.

Synaptomys sp.—bog lemming

Site record (1).—Frontier Co.: Lime Creek Site, 25FT41, north side of Lime Creek valley, 1 mi. from original junction of Lime and Medicine creeks (Graham et al., 1987: 283).

Remarks.—This record probably applies to the southern bog lemming because there are only two species in the genus *Synaptomys*. The second species is the northern bog lemming, *S. borealis*, but the nearest modern records to Nebraska are from northern Minnesota (Hall, 1981). However, Pleistocene fossils of *S. borealis* have been found in Kansas and Tennessee (Linzey, 1983).

This Holocene record from the Lime Creek site comes from a place that is geographically very near the Frontier Co. record of *S. cooperi* at the Allen site. The Lime Creek site is also of interest because it is an early Holocene site falling in the range of 9500 to 8000 years ago. This record helps establish that *Synaptomys* has had an early and relatively long history in southwestern Nebraska and that the Republican River was the route followed to form the isolated population in Dundy Co.

Oryzomys texensis—Texas rice rat

Site records (10).—Boone Co.: Beaver Creek Site, 25BO23, 2.4 km N Loretto (Koch, 2002: i, 1-4; Koch and Nelson, 2002: 87). Butler Co.: Palmer Johnson Site, 25BU37, east of Ulysses (Souders, 1994: 53; Bozell and Koch, n.d.). Douglas Co.: Parker Site, 25DO2, on the bluffs north of Ponca Creek and west of the Missouri River, in extreme northeastern Douglas Co. (Ewing, 2000: 5). Frontier Co.: 25FT22, Sec. 11, T5N, R26W (Turnmire, 1996: 189-191) ?. Howard Co.: Schmidt Site, 25HW301, along North Loup River, near Elba (Satorius-Fox, 1982: 8; Graham et al., 1987: 284). Nance Co.: 25NC10, vicinity of Fullerton (Graham et al., 1987: 284). Richardson Co.: Rulo Southeast, 25RH69, 0.4 km S Rulo (Nelson, 2006: 165); Rulo Southeast, 25RH70, 0.6 km S Rulo (Nelson, 2006: 170, 173). Sarpy Co.: Patterson Site, 25SY31, ≈ 7 mi. S Gretna (Bozell and Ludwickson, 1999: 83). Webster Co.: Shipman Site, 25WT7, south side of Republican River, between Red Cloud and Guide Rock (Graham et al., 1987: 286).

Remarks.—At least 10 sites in Nebraska have been reported having remains of the Texas rice rat. This species is not part of the modern mammalian fauna of Nebraska. Modern rice rats are essentially obligate mesic habitat dwellers confining their activities to areas along rivers, streams, and lakes, especially in the drier portions of their geographic range. These records in Nebraska occur in several river systems—three along the Missouri, two in the Republican drainage, one along the main stem of the Platte, three in the Loup system, which enters the Platte near Columbus, and one in the drainage of the Big Blue. The latter site is in the uppermost reaches of the Big Blue River drainage and probably resulted from crossing over from the nearby Platte River drainage.

Based on the occupation times of these sites, it appears that rice rat populations entered Nebraska between AD 1000 to 1100 first appearing along the Missouri River and the lower Platte River. The initial occupation times for the villages further upstream along the Big Blue and Loup rivers fall between AD 1100 and AD 1275. The populations along the Republican River may have reached the areas in Nebraska via the Kansas River, which joins the Missouri River near

Kansas City. The populations along the Republican River appear to have arrived post-AD 1350, later than other Nebraska populations. The last evidence of rice rat occurrence in Nebraska falls around AD 1550, which is in line with what was happening in other parts of the geographic range of rice rats.

The site in Boone County is the northernmost for the species. There were remains of a minimum of 11 individuals recovered at this site (Koch and Nelson, 2002) and at the Schmidt site in Howard Co., which is the western-most in the Platte drainage, remains of 37 individuals were recovered (Satorius-Fox, 1982; Graham et al., 1987). If the record in Frontier Co. proves to be correct, it would be the western-most for *O. texensis* (Turnmire, 1996).

***Sigmodon hispidus*—cotton rat**

Site record (1).—Nance Co.: Palmer Locality [in part], 25NC29, 5.5 mi. S, 19 mi. W Fullerton (Meadow and Peterson, 2001: 157).

Remarks.—This is a record that I must question. The northward dispersal of this southern species over the last century or more has been closely monitored. The first record from Kansas was in 1892, with animals dispersing from Oklahoma (Cockrum, 1948). The species reached Richardson Co. in southeastern Nebraska by 1958 (Jones, 1960). Its initial spread northward and then to the west in Nebraska has been documented by Choate and Genoways (1967), Genoways and Schlitter (1967), Farney (1975), and Wright et al. (2010). The northern-most of the modern records for this cotton rat are in Hayes and Kearney cos. The Holocene record from Nance Co. is about 112 km [70 mi.] to the northeast of the northern record from just south of Platte River south of Kearney in Kearny Co. There is no other evidence of cotton rats on the Great Plains during the Holocene (Graham et al., 1987). Instead of representing the hispid cotton rat, I believe that this is probably a Texas rice rat, *Oryzomys texensis*. These two species are the only representatives of rodent subfamily Sigmodontinae in Nebraska so they are relatively closely related. In addition as seen above, there is a record of *Oryzomys texensis* known from Nance Co. at site 25NC10 near Fullerton.

***Neotoma floridana*—eastern woodrat**

Site records (3).—Frontier Co.: Mowry Bluff Site, 25FT35, 6 mi. NW Cambridge (Falk, 1969b: 47; Mick, 1983: 205; Graham et al., 1987: 286); Allen Site, 25FT50, ≈ 3 mi. upstream from Medicine Creek Dam (Hudson, 2007: 195). Garden Co.: Ash Hollow Cave Site, 25GD2, 3 mi. SE Lewellen (Champe, 1946: 43-44).

Remarks.—The eastern woodrat occurs in southwestern Nebraska south of the Platte River, with the northern-most record being North Platte in Lincoln Co. There is also an isolated population along the Niobrara River in north-central Nebraska. Recently, Benedict et al. (2008b) discovered a third population of eastern woodrats in southern Gage and Jefferson cos. The preferred habitat of the eastern woodrat includes woody vegetation, which would provide some of the material that could be used to construct stick "houses." These "houses" provide protection to the nest chamber and food stores. Besides sticks, woodrat "house" construction would have included many other items from the immediate environs of the "house," such as cactus pads, nuts, animal dung, and human refuses.

The Mowry Bluff and Allen Holocene sites in Frontier Co. are within the modern distribution of the eastern woodrat in Nebraska. On the other hand, the Ash Hollow Cave site is approximately 112 km [70 mi.] west-northwest of the nearest modern record at North Platte. The Holocene record from the Allen site places eastern woodrats in this portion of Nebraska as early as 9500 to 13,000 years ago. The Mowry Bluff site puts the eastern woodrat in Frontier Co. in southwestern Nebraska about 1000 years ago. The date of the Ash Hollow Cave site remains are less precisely known but are from the last 2100 years. These data seem sufficient to recognize that *N. floridana* has been part of the fauna of southwestern Nebraska throughout the Holocene.

Neotoma sp.—woodrat

Site records (4).—Frontier Co.: 25FT13, 16, 17, vicinity of Medicine Creek Dam (Mick, 1983: 174, 184, 190; Graham et al., 1987: 284). Webster Co.: Shipman Site, 25WT7, south side of Republican River, between Red Cloud and Guide Rock (Mick, 1983: 153; Graham et al., 1987: 286).

Remarks.—There are currently two species of woodrats occurring in Nebraska, as the bushy-tailed woodrat, *Neotoma cinerea*, is present as well as the eastern woodrat. The bushy-tailed woodrat is confined to the western part of the Nebraska Panhandle with a record from Oshkosh, Garden Co., being the furthest east modern record (Jones, 1964; Genoways et al., 2008b).

The Holocene woodrat sites listed for *Neotoma* sp. all fall in the modern geographic range of the eastern woodrat, *N. floridana*. All of these records date to the last 1100 years.

Peromyscus leucopus—white-footed mouse

Site record (1).—Frontier Co.: 25FT17, vicinity of Medicine Creek Dam (Mick, 1983: 190; Graham et al., 1987: 286; Kivett and Metcalf, 1997: 213).

Remarks.—The white-footed mouse is one of the common small mammals in Nebraska. It occurs over the eastern two-thirds of the state and is expanding its geographic range westward along the major river systems. Along the Republican River in work by Jones (1964), the furthest west record was McCook, Red Willow Co., based on specimens that were trapped in 1941. Benedict et al. (2000) reported on the results of a 1990 survey along the Republican River. At that time the westernmost specimen was captured at 0.5 mile south and 6 miles west of Parks, Dundy County, which was approximately 100 km [62 mi.] west of McCook. Benedict et al. (2000) postulated that the westward movement of this species was the result of development of riparian woodlands along the river providing the preferred wooded habitat of the white-footed mouse. This species readily climbs trees in search of its food of seeds, fruits, buds, and other plant materials.

There is only a single Holocene record of this species in Nebraska and it is based on the remains of a single individual (Graham et al., 1987). This one record does document the occurrence of *P. leucopus* along one of the tributaries—Medicine Creek—of the Republican River about as far west as McCook at some point between 700 and 900 years ago. Whether or not the white-footed mouse occurred in this area continuously from that date forward is not known but clearly it, at least, reached this area before modern time.

Peromyscus maniculatus—deer mouse

Site records (4).—Brown Co.: McIntosh Site, 25BW15, Enders Lake, ≈12 mi. S, 12.5 mi. W Ainsworth (Koch, 1995: 43; 2004: 117). Frontier Co.: Mowry Bluff Site, 25FT35, 6 mi. NW Cambridge (Falk, 1969b: 47; Mick, 1983: 205; Graham et al., 1987: 286). Hall Co.: Hulme Site, 25HL28, 16 mi. W Grand Island (Bozell, 1991: 233). Howard Co.: Schmidt Site, 25HW301, along North Loup River, near Elba (Satorius-Fox, 1982: 8; Mick, 1983: 155; Graham et al., 1987: 284).

Remarks.—The deer mouse is the commonest small mammal in Nebraska, occurring throughout the state. It is present in most habitats even in wooded areas when the white-footed mouse is not present. The deer mouse does not climb for its food.

The four Holocene sites for the deer mouse in Nebraska are scattered in central to southwestern Nebraska. These deposits were all laid down in the time period of 400 to 1000 years ago.

Peromyscus **sp.—mouse**

Site records (7).—Boone Co.: Beaver Creek Site, 25BO23, 2.4 km N Loretto (Koch, 2002: i, 1-4; Koch and Nelson, 2002: 87). Douglas Co.: Cabanne's Post, 25DO8, near Dodge Park northeast of Omaha (Bozell et al., 1990: 32). Frontier Co.: 25FT22, Sec. 11, T5N, R26W (Turnmire, 1996: 189-191); Lime Creek Site, 25FT41, north side of Lime Creek valley, 1 mi. from original junction of Lime and Medicine creeks (Graham et al., 1987: 283). Harlan Co.: Stevenson Village, 25HN31, 1.5 mi. S, 0.25 mi. W Alma (Adair and Brown, 1987: 123, 574). Nance Co.: Palmer Locality [in part], 25NC29, 5.5 mi. S, 19 mi. W Fullerton (Meadow and Peterson, 2001: 157). Sarpy Co.: Patterson Site, 25SY31, ≈ 7 mi. S Gretna (Bozell and Ludwickson, 1999: 83).

Remarks.—These records all pertain to either to *P. leucopus* or *P. maniculatus* because there are no other species of the genus near to Nebraska. All of these records come from parts of the state where both species should be occurring. The ages of all but one of these Holocene sites fall in the range of the *P. leucopus/P. maniculatus* sites above. The one exception is the Lime Creek site (25FT41) where the recovered material falls into the time period of 8000 to 9500 years ago. This gives a far earlier documented presence of *Peromyscus* in Nebraska than other sites

Reithodontomys **sp.—harvest mice**

Site records (5).—Boone Co.: Beaver Creek Site, 25BO23, 2.4 km N Loretto (Koch, 2002: i, 1-4; Koch and Nelson, 2002: 87). Colfax Co.: Schuyler Site, 25CX1, along Shell Creek, ≈ 3 mi. NW Schuyler (Bozell et al., 1982: 27; Graham et al., 1987: 286). Frontier Co.: 25FT22, Sec. 11, T5N, R26W (Turnmire, 1996: 189-191). Howard Co.: Schmidt Site, 25HW301, along North Loup River, near Elba (Satorius-Fox, 1982: 8; Mick, 1983: 155; Graham et al., 1987: 284). Sarpy Co.: Patterson Site, 25SY31, ≈ 7 mi. S Gretna (Bozell and Ludwickson, 1999: 83).

Remarks.—There are two species of harvest mice occurring in Nebraska, with both occurring essentially statewide. The western harvest mouse, *R. megalotis*, inhabits grassy areas throughout the state but seems to prefer dense growths of grass associated with moister habitats. The plains harvest mouse, *R. montanus*, prefers drier upland grassland and may be excluded from forested areas along the Missouri River. These species are relatively easy to distinguish based on external characteristics, but they are far more difficult to separate based on skeletal material, especially if it is fragmentary. The remains from these five Holocene sites in eastern, east-central, and southwestern Nebraska could be one or the other of the species of harvest mice or both. The time period covered by these sites is about 300 to 1000 years ago.

Onychomys leucogaster—northern grasshopper mouse

Site records (9).—Boone Co.: Beaver Creek Site, 25BO23, 2.4 km N Loretto (Koch, 2002: i, 1-4; Koch and Nelson, 2002: 87). Brown Co.: McIntosh Site, 25BW15, Enders Lake, ≈12 mi. S, 12.5 mi. W Ainsworth (Koch, 1995: 43; 2004: 117). Butler Co.: Palmer Johnson Site, 25BU37, east of Ulysses (Souders, 1994: 53; Bozell and Koch, n.d.). Frontier Co.: 25FT14, vicinity of Medicine Creek Dam (Kivett and Metcalf, 1997: 213); 25FT22, Sec. 11, T5N, R26W (Turnmire, 1996: 189-191); Mowry Bluff Site, 25FT35, 6 mi. NW Cambridge (Falk, 1969b: 47; Mick, 1983: 205; Graham et al., 1987: 286). Gosper Co.: Wallace Site, 25GO2, Plum Creek, 5.5 mi. N, 9.5 mi. E Elwood (Winfrey, 1991: 82; Souders, 1994: 53). Howard Co.: Schmidt Site, 25HW301, along North Loup River, near Elba (Satorius-Fox, 1982: 8; Mick, 1983: 156; Graham et al., 1987: 284). Sherman Co.: Bill Packer Site, 25SM9, along Davis Creek in extreme northeastern corner of county (Graham et al., 1987: 284; Bozell and Rogers, 1989: 27; Souders, 1994: 53).

Remarks.—The northern grasshopper mouse occurs throughout most of Nebraska, with the possible exception of counties bordering the Missouri River north of Omaha (Jones,1964; Benedict et al., 2000; Genoways et al., 2008b). The species is abundant in western Nebraska becoming rare in the east. The species prefers grassland habitats, which probably accounts for it being rare to absent in the tall grass and riparian habitats along the Missouri River. The northern grasshopper mouse is carnivorous in its dietary habits, whenever arthropods, other invertebrates, and small vertebrates, are available in the environment.

The Holocene sites for the northern grasshopper mouse in Nebraska are missing from that part of the state—western—where they are abundant in modern times, with the western-most records from Frontier Co. The records are concentrated in four counties in east-central Nebraska and two counties in southwestern Nebraska. There is a single site in the Sand Hills in Brown Co. The times of origin of these Holocene sites are encompassed by 400 to 1500 years ago.

Rattus norvegicus—Norway rat

Site record (1).—Sarpy Co.: Lucien Fontenelle's Post, 25SY26, Bellevue (Bozell et al., 1990: 29).

Remarks.—The Norway rat is an invasive European species belonging to the Old World family of rats and mice—Muridae. Silver (1927) believed that the Norway rats had been "introduced into the United States, probably from England, about the first year of the American Revolution, 1775." However, recent research along the east coast has shown that there were earlier, possibly isolated, populations in such places as Fort Frederica, GA, in 1736 to 1750 (Reitz and Honerkamp, 1983) and Stobo Plantation, SC, in 1720-1740 (Webber and Reitz, 1999). Lucien Fontenelle's Post was operated under various ownerships from about 1822 to 1842. Fontenelle was from a New Orleans family who acquired this post at the site of the future Bellevue, Sarpy Co., in 1825. This is a very early record for the Norway rat in the central United States. The remains represent a minimum of 53 individuals, indicating that there was an established population present. These rats undoubtedly reached the post via steamboat from St. Louis. The sources of the St. Louis population could have been either Pittsburgh, PA, along the Ohio River and/or New Orleans, LA, downstream on the Mississippi River. The first steamboat to reach this area of eastern Nebraska was the Western Engineer with the Long Expedition in 1819 (Genoways and Ratcliffe, 2008).

Erethizon dorsatum—common porcupine

Site records (4).—Dakota Co.: Big Village Site, 25DK5, 1 mi. NE Homer (O'Shea and Ludwickson, 1992: 331). Frontier Co.: 25FT13, vicinity of Medicine Creek Dam (Mick, 1983: 174; Graham et al., 1987: 284; Kivett and Metcalf, 1997: 213); Allen Site, 25FT50, ≈ 3 mi. upstream from Medicine Creek Dam (Hudson, 2007: 195). Morrill Co.: Greenwood Site, old Greenwood Stage Station, Keenan Ranch, ≈ 9 mi. S, 8 mi. E Redington (Renaud, 1933: 14).

Remarks.—Jones (1964; Genoways et al., 2008a) considered the common porcupine in Nebraska to be statewide in distribution, but it probably was never common in the eastern third of the state. This herbivore is the second largest rodent in Nebraska being outweighed only by the American beaver. It consumes a variety of green plants, twigs, and leaves in the growing season and relies on bark, especially the cambium layer, during the winter months.

The four Holocene sites certainly speak to the statewide distribution of the species in Nebraska—Dakota Co. in the northeast, Frontier Co. in the southwest, and Morrill Co. in the central Panhandle. The sites cover the time period of the Holocene as well, with the Allen site being oldest at 9500 to 13,000 years old. The other record from Frontier Co. was from about 1000 years ago and the Dakota and Morrill cos. sites were quite recent probably being from the last 300 years.

Porcupines are considered to be one of the best survival foods available so it is surprising that their remains have been found in only four Holocene archaeological sites. Porcupines are easy to capture and kill just using a stick and can be eaten raw in emergency situations.

DISCUSSION

Information in the preceding species accounts give some documentation on a minimum of 57 species of mammals (with a potential of eight additional species) during the Holocene in Nebraska. Most of these species remained part of the mammalian fauna of Nebraska until 1850. After this point many of these species, especially the larger species, were extirpated from the mammalian fauna of Nebraska directly through human activities and indirectly through habitat loss and environmental change.

The most interesting occurrence in the Holocene record of Nebraska mammals was the appearance of the Texas rice rat, *Oryzomys texensis*, about 1000 years ago. Over the next 500 years the species moved westward along the river systems of Nebraska into Boone, Frontier, and Howard cos. Then as suddenly as they arrived the species disappeared from the fossil record around AD 1550. The occurrence of these mice in the state required the physical movement of individual rice rats along the river systems. On the other hand, the disappearance of the species did not require any movement, but only a series of extinction events. The movement of the species into the state appears to have happened in a rather sequential manner proceeding up each river system, but the extinction events would have occurred in a more random manner. Those populations in more favorable habitats would have held out longer than those with less resources available.

The phenomenon of the extensive range expansion by the Texas rice rat is a complicated topic, which will be the subject of a separate publication. The Nebraska part of story of the Texas rice rat is only a small portion of a much larger event that impacted the upper Mississippi River basin from the mouth of the Ohio River northward. Beginning about 2500 years ago, Texas rice rat populations from a center in southern Illinois, western Kentucky, and adjacent parts of Missouri and Arkansas exploded up the Mississippi, Illinois, Ohio, and Missouri river systems. Along the Illinois River rice rats reached northward nearly to Peoria and along the Ohio River populations reached eastward into the western third of Pennsylvania. Finally, along the Missouri River system rice rats reached into Kansas, Iowa, and Nebraska. It is my contention that this event started when native people began to settle into more permanent villages and began to undertake gardening, using several eastern North American plants as well as squash and gourds rather than relying strictly on a hunter-gatherer subsistence. The expansion was slow at first, but with the addition of maize (*Zea mays*) to the crops being grown around AD 900 there was rapid expansion of the rice rat populations reaching their maximum expansion. The history of these rice rats was tightly associated with native people where they acted as a commensal pest of stored food stuff and sought shelter in native house structures. The collapse of this expansion begins with the full impact of the Little Ice Age beginning around AD 1400. The temperatures in parts of North America dropped 1° to 2° C, but the larger problem for the native people as well as the rice rats were periods of extended drought causing the failure of the maize crop on which they depended as the main part of their diets.

For five species that remain in Nebraska mammalian fauna, there are records from southwestern Nebraska that document some shifts in their geographic ranges during the Holocene. Two species, *Urocitellus elegans* and *Thomomys talpoides*, had geographic ranges that have retracted to the west so both have recent records only from the western Panhandle. Hoffmann and Jones (1970) classified Recent populations of these two species as having montane affinities. This recalls that a significant number of the species in the Pleistocene North Cove fauna had montane-boreal relationships moving west and north. Records from Frontier and Red Willow cos. document that Audubon's cottontail, *Sylvilagus audubonii*, occurred at least 100 km further to the east in the last 2000 years than they have been documented in modern times (Jones, 1964). Hoffmann and Jones (1970) placed the affinities of this species in the southwestern United States. Species with southwestern origins have a center of distribution in the area of the northern edge of Chihuahuan-Sonoran desserts. This small retraction to the west should be interpreted as readjustment of the geographic range shifting toward this center of distribution.

The final two species, *Lepus townsendii* and *Urocyon cinereoargenteus*, have retracted their geographic ranges to the north. The retreat northward of the white-tailed jackrabbit has been documented in Kansas and in Nebraska during the post-1850 time period. The species was abundant in western Kansas prior to European settlement, but by the late 1940s the species was only present in the extreme northeastern part of the state (Carter, 1939; Brown, 1947). A similar situation prevailed in Nebraska where Holocene records place the species in Frontier Co., but by the 1950s the species no longer occurred south of the Platte River (Jones, 1964). This retreat may be continuing. Several theories have been proposed for this retreat, but conversion of the prairie to row crops and a warming climate have been the most often cited. The white-tailed jackrabbit was classified by Hoffmann and Jones (1970) as Steppe species with similar species having a center of distribution in northern Nebraska and South Dakota. The gray fox has an unusual history because it is a tropical species that has an affinity to the eastern deciduous forest entering Nebraska from the east (Hoffmann and Jones, 1970).

Another species with a history similar to the gray fox is the Virginia opossum, *Didelphis virginiana*, a member of the modern mammalian fauna of Nebraska, which is absent from the Holocene record. I find this to be the most surprising absence from the Holocene record because the Virginia opossum is a medium-sized mammal, associated with the deciduous forests, and is still hunted as a game species in Nebraska. I doubt that native people would have bypassed this source of protein if it was readily available. This raises the potential that the Virginia opossum occupied a very narrow range confined to the Missouri River lowlands in eastern Nebraska during the Holocene. This species may have extended its geographic range quickly to the west as deciduous forest developed along watercourses with the suppression of prairie fires. Three other species (*Sciurus niger*, *S. carolinensis*, and *Marmota monax*) with affinities to eastern deciduous forests are represented in the Nebraska Holocene by only a couple of scattered records in improbable places. These species would be most expected in forests along the Missouri River and would be expected to form at least a small portion of the diet of native people living in this area. All of this evidence would seem to be pointing to the fact that the deciduous forest along the Missouri was not well developed and populations of these species adapted to this habitat were rare at best.

Two species of mammals have questionable Holocene evidence from Nebraska. The moose, *Alces alces*, is represented by a tool made from an antler tine. This could have been a traded item or from a wandering individual that entered Nebraska. Solitary wandering moose, mostly if not all young males, are regularly present currently in the state. The hispid cotton rat, *Sigmodon hispidus*, is represented by a single Holocene record, which I believe is an incorrectly identified specimen. This species did not enter Nebraska prior to the mid-twentieth century (Jones, 1964).

There are no Holocene records of the nine-banded armadillo, *Dasypus novemcinctus*, from Nebraska, but there are records of the late Pleistocene armadillo, *D. bellus*, from the state (Voorhies, 1987). This has led to some confusion about the status of the armadillo in Nebraska because the fossil and living species of armadillo are closely related morphologically. The first modern nine-banded armadillo reached the state in the early 1970s and the species has been documented progressively further north in the state. The individuals entering Nebraska are males and the nearest breeding population is currently in central Kansas (Freeman and Genoways, 1998).

There are some mammalian species that are not present in the Holocene record of Nebraska that probably are simply the result of sampling error. These are species with small individuals that would not be expected in an archeological site except as an accidental occurrence. These groups include bats (order Chiroptera), shrews of the genus *Sorex*, and some of the smaller rodents. Two of the smallest carnivores—*Mustela nivalis* and *Spilogale putorius*—which are currently statewide in distribution, but are rare in numbers, are missing from the Holocene record. Because these species absences can be predicted primarily on the basis of size, they should not be considered of any biological significance.

The take away message here is that there were extensive shifts at the end of the Pleistocene in both the composition and distribution of mammalian species. In the early Holocene the compression of the mammalian fauna was relieved and species shifted their geographic ranges with the changing climate. This was followed by a period of relative stability in Nebraska's mammalian fauna as shown by the Holocene records. There were no extinct species represented in the mammalian Holocene record of Nebraska. With a few exceptions as discussed above, Holocene species remained a part of the modern mammalian fauna of the state. There was shifting of geographic ranges of some species during the mid to late Holocene but most of these were on a small scale of a few 100 km or less. The one dramatic exception to this rather stable picture was the invasion and then collapse of the Texas rice rat population in the state. This 500-year occurrence in Nebraska was unique among Holocene mammals and requires additional study to understand its basis. However, with the beginning of European activities and settlement the Holocene picture of stability in Nebraska's mammalian fauna is shattered with changes rivaling those at the end of the Pleistocene. In the time period post-1850, the top predators in the state—gray wolf, mountain lion, black bear, grizzly bear, North American river otter, and black-footed ferret—were extirpated from the state or became extinct. Much the same thing occurred with the large herbivores—bison, mule deer, white-tailed deer, elk, pronghorn, and bighorn sheep. By the early twentieth century these populations were so decimated that some species such as the white-tailed deer, mule deer, and pronghorn were brought under intensive game management and the populations allowed to rebuild. At the end of the twentieth century and beginning of the twenty-first century, elk populations were rebounding as were some carnivore populations under protection of the federal Endangered Species Act (Hoffman and Genoways, 2005). Much the same history was shared by the beaver population, which is still regaining some of its ancient homeland. Not all impacts on species post-1850 have been negative. The nine-banded armadillo and hispid cotton rat have expanded their geographic ranges into the state potentially because of a warming climate. Species with eastern deciduous forest affinities—Virginia opossum, eastern fox squirrel, woodchuck, and white-footed deermouse—have expanded their geographic ranges westward along with riparian forests as prairie fires and grazing by large herbivores were limited. However, it must be remembered that these gains in geographic range by forest and forest-edge species has resulted in the loss of space by species that are prairie adapted. The shifting geographic ranges and changes in faunal composition is certain to continue into the future given the impact of climate change in the twenty-first century.

Acknowledgments

This publication would not have been possible without the time and cooperation of many people with responsibility for the preservation of paleontology and archeology site reports and publications. I wish to particularly acknowledge Kelli Bacon, Trisha Nelson, and Rob Bozell of the Archeology Division, Nebraska State Historical Society, and Alan Osborn, Curator of Anthropology, University of Nebraska State Museum. The Interlibrary Librarians at Love Library, University of Nebraska-Lincoln, were indefatigable in their efforts to obtain copies of literature held by other institutions. I particularly appreciate the time and support of Kelli Bacon and Rob Bozell, Nebraska State Historical Society, Lincoln, George Corner, University of Nebraska State Museum, Lincoln, and Justin Hoffman, McNeese State University, Lake Charles, Louisiana, for reading and commenting on a draft of this manuscript. I also appreciate the opportunity created by Paul Royster and Sue Ann Gardner so this research could be published in Zea eBooks. A special thank you to my wife, Joyce, who took on the thankless task of serving as my copy editor. Ted Genoways kindly prepared the design for the front cover. The Nebraska map on the front cover is used royalty free courtesy of Bruce Jones Design, 2011. The map on the inside of the back cover was prepared by Angie Fox, scientific illustrator, University of Nebraska State Museum.

Literature Cited

Abramov, A.V. 2000. A taxonomic review of the genus *Mustela* (Mammalia, Carnivora). Zoosystematica Rossica, 8(2), 357–364.

Adair, M. J. (ed.). 1989. Archaeological investigations at the North Cove Site, Harlan County Lake, Harlan County, Nebraska. Kaw Valley Engineering and Development, Inc., Junction City, KS, vi + 1-106.

Adair, M. J., and K. L. Brown (eds.). 1987. Prehistoric and historic cultural resources of selected sites at Harlan County Lake, Harlan County, Nebraska: test excavations and determination of significance for 28 sites. Kaw Valley Engineering and Development, Inc., Junction City, KS, xiii + 594 pp.

Agenbroad, L. D. 1977. Climatic change and early man in northwest Nebraska. Museum Journal, 17: 117-125.

Agenbroad, L. D. 1978. The Hudson-Meng Site: An Alberta bison kill in the Nebraska High Plains. Washington, DC: University Press of America, xvi + 230 pp.

Aubry, K. B., M. J. Statham, B. N. Sack, J. D. Perrine, and S. M. Wisely. 2009. Phylogeography of the North American red fox: vicariance in Pleistocene forest refugia. Molecular Ecology, 18: 2668-2686.

Bamforth, D. B. 2002. The Paleoindian occupation of Medicine Creek drainage, southwestern Nebraska. Pp. 54-83, *in* Medicine Creek: Seventy years of archaeological investigations, ed. D. C. Roper. Tuscaloosa: University of Alabama Press, xix + 248 pp.

Bamforth, D. B. 2007. Introduction. Pp. 1-8, *in* The Allen Site: a Paleoindian camp in southwestern Nebraska, ed. D. B. Bamforth. Albuquerque: University of New Mexico Press, xvi + 284 pp.

Bartlein, P. J., T. Webb III, and E. Fleri. 1984. Holocene climatic change in the northern Midwest: pollen-derived estimates. Quaternary Research, 22: 361-374.

Barton, H. D., and S. M. Wisely. 2012. Phylogeography of striped skunks (*Mephitis mephitis*) in North America: Pleistocene dispersal and contemporary population structure. Journal of Mammalogy, 93: 38-51.

Beauvais, G. P., and L. Johnson. 2004. Species assessment for wolverine (*Gulo gulo*) in Wyoming. Prepared for United States Department of the Interior, Bureau of Land Management, Wyoming State Office, Cheyenne, WY, 46 pp.

Bell, E. H. (ed.). 1936. Chapters in Nebraska archaeology. Lincoln: University of Nebraska, 427 pp.

Bell, E. H., and R. E. Cape. 1936. The rock shelters of western Nebraska in the vicinity of Dalton, Nebraska. Pp. 357-399, *in* Chapters in Nebraska archaeology, ed. E. H. Bell. Lincoln: University of Nebraska, 427 pp.

Bell, E. H., and G. H. Gilmore. 1936. The Nehawka and Table Rock foci of the Nebraska Aspect. Pp. 301-355, *in* Chapters in Nebraska archaeology, ed. E. H. Bell. Lincoln: University of Nebraska, 427 pp.

Benedict, R. A. 1999a. Morphological and mitochondrial DNA variation in a hybrid zone between short-tailed shrews (*Blarina*) in Nebraska. Journal of Mammalogy, 80:112-134.

Benedict, R. A. 1999b. Characteristics of a hybrid zone between two species of short-tailed shrews (*Blarina*). Journal of Mammalogy, 80:135-141.

Benedict, R. A., H. H. Genoways, and P. W. Freeman. 2000. Shifting distributional patterns of mammals in Nebraska. Transactions of the Nebraska Academy of Sciences, 26: 55–84.

Blackman, E. E. 1907. Report of archeologist. Nebraska State Historical Society Publications, 15: 323-359.

Blackman, E. E. 1924. Exploration of aboriginal remains in Loup Valley. Nebraska History, 7: 1-8.

Blakeslee, D. J. 1988. An introduction to St. Helena archaeology. Pp. 1-12, *in* St. Helena archaeology: new data, fresh interpretations, ed. D. J. Blakeslee. Reprints in Anthropology, 39: vi + 1-193.

Blakeslee, D. J., and W. W. Caldwell. 1979. The Nebraska Phase: an appraisal. Reprint in Anthropology, 18: viii + 1-186.

Blakeslee, D. J., and J. O'Shea. 1983. The Gorge of the Missouri: an archeological survey of Lewis and Clark Lake, Nebraska and South Dakota. Archaeology Laboratory, Wichita State University, Wichita, KS, 379 pp.

Bleed, P. 1973. A report on 1973 archaeological investigations in the Wood River Valley, Buffalo County, Nebraska. Midstate Archaeological Survey, National Park Service, v + 122 pp.

Bozell, J. R. 1988. Changes in the role of the dog in protohistoric-historic Pawnee culture. Plains Anthropologist, 33: 95-111.

Bozell, J. R. 1991. Fauna from the Hulme Site and comments on Central Plains Tradition subsistence variability. Plains Anthropologist, 36: 229-253.

Bozell, J. R. 1993. Vertebrate remains from 1991 archeological investigations at Agate Fossil Beds National Monument, Sioux County, Nebraska. Pp. 59-82, *in* Archeological survey and testing at Agate Fossil Beds National Monument, Sioux County, Nebraska, ed. C. P. Clark. National Park Service, Midwest Archeological Center, Technical Report, 22: vi + 1-82.

Bozell, J. R. 1994a. Vertebrate remains from select 1991 and 1992 archeological investigations at Agate Fossil Beds National Monument, Sioux County, Nebraska. Pp. 45-63, *in* Archeological investigations on Marsland Road, Agate Fossil Beds National Monument, Sioux County, Nebraska, ed. C. P. Clark. National Park Service, Midwest Archeological Center, Technical Report, 31: v +1-63.

Bozell, J. R. 1994b. Faunal remains from 1992 archeological investigations at Scotts Bluff National Monument, Scotts Bluff County, Nebraska. Pp. 49-66, *in* Archeological survey of Scotts Bluff National Monument, Scotts Bluff County, Nebraska, ed. C. P. Clark. National Park Service, Midwest Archeological Center, Occasional Studies in Anthropology, 31: vi + 1-66.

Bozell, J. R. 1995. Culture, environment, and bison populations on the late prehistoric and early historic central plains. Plains Anthropologist, 40: 145-163.

Bozell, J. R., A. A. Buhta, R. D. Mandel, S. R. Holen, and E. J. Lueck. 2008. An archeological and geomorphic survey of selected lands in Broadwater West and Blue Creek study areas, Morrill and Garden counties, Nebraska. Archeology Laboratory, Augustana College, Sioux Falls, SD, Archeological Contract Series, 226: xiv +1-254.

Bozell, J. R., S. Donnelly, M. Rogers, and S. Zink. 1990. Animal utilization at two nineteenth century trading posts in east central Nebraska. Report on file at the Nebraska State Historical Society, Lincoln, vi + 125 pp.

Bozell, R., and A. Koch. n.d. Faunal remains from the Palmer Johnson site (25BU37), Butler County, Nebraska. Unpublished manuscript deposited at the Nebraska State Historical Society, Lincoln, 16 pp.

Bozell, J. R., and J. Ludwickson. 1988. Highway archeological investigations at the Slaughterhouse Creek Site and other cultural resources in the Pine Ridge area. Nebraska State Historical Society, Report for the Nebraska Department of Roads, vi + 102 pp.

Bozell, J. R., and J. Ludwickson. 1994. Little Pawnee Creek Site: physical setting and investigation strategy. Pp. 100-107, *in* Nebraska phase archeology in the South Bend locality, eds. J. R. Bozell and J. Ludwickson. Nebraska State Historical Society, Report for Nebraska Department of Roads, xiii + 236 pp.

Bozell, J. R., and J. Ludwickson. 1999. Archeology of the Patterson Site: Native American life in the lower Platte Valley, A. D. 1000-1300. Report by the Nebraska State Historical Society to the Nebraska Department of Roads and Federal Highway Administration, viii + 210 pp.

Bozell, J. R., and R. E. Pepperl. 1982. Emergency evaluation of Native American archeological resources encountered during construction of Section 1, Mirdan Canal, Garfield County, Nebraska. Division of Archeological Research, Department of Anthropology, University of Nebraska-Lincoln, Technical Report No. 82-11: iv + 1-24.

Bozell, J. R., and R. E. Pepperl. 1987. A cultural resources study of the Crow Butte Uranium Prospect, Dawes County, Nebraska. Nebraska State Historical Society, Report for Resources Technologies Group, Inc., vi + 83 pp.

Bozell, J. R., and M. K. Rogers. 1989. A Great Oasis fauna from central Nebraska. Central Plains Archaeology, 1: 3-36.

Bozell, J. R., L. M. Snyder, and C. R. Falk. 1982. Descriptive analysis of unmodified vertebrate remains recovered from the Schuyler Site (25CX1), Colfax County, Nebraska. Division of Archeological Research, Department of Anthropology, University of Nebraska-Lincoln, Technical Report No. 82-06, iii + 1-110.

Brooking, A. M. 1921. Ancient house sites at Meadow, Nebraska. Nebraska History, 4: 37-39.

Brown, H. L. 1947. Why has the white-tailed jack rabbit (*Lepus townsendii campanius* Hollister) become scarce in Kansas? Transactions of the Kansas Academy of Science, 49: 455-456.

Buhta, A. A., J. M. Kruse, and J. R. Bozell. 2007. An intensive cultural resources inventory of the upper and Little Blue rivers drainage systems: archeological and geomorphological reconnaissances in Adams, Antelope, Clay, Holt, Nuckolls, and Thayer counties, Nebraska. Volume 1: Cultural resources report. Archeology Laboratory, Augustana College, Sioux Falls, SD, report to the Nebraska State Historical Society, xx + 367 pp.

Butler, T. 1994. Lithic analysis of surface materials at 25SX163. Pp. 29-44, *in* Archeological investigations on Marsland Road, Agate Fossil Beds National Monument, Sioux County, Nebraska, ed. C. P. Clark. National Park Service, Midwest Archeological Center, Technical Report, 31: v +1-63.

Cannon, K. P. 1991. Faunal remains from site 25SF50. Pp. 57-72, *in* Test excavation at site 25SF50, Scotts Bluff National Monument, Nebraska, ed. K Griffin. National Park Service, Midwest Archeological Center, Technical Report, 5: vi + 1-83.

Carlson, G. F. 1973. Archeological salvage and survey in Nebraska: Part I: Highway archeological and historical salvage investigations in Nebraska, 1965 to 1968. Nebraska State Historical Society, Publications in Anthropology, 5: xiv + 1-155.

Carlson, G. F. 1976. Archeological salvage investigations at Santee, Nebraska. Report on file at the Nebraska State Historical Society, Lincoln, 7 pp.

Carlson, G. F. 1979. Archeological investigations at Fort Atkinson (25WN9), Washington County, Nebraska, 1956-1971. Nebraska State Historical Society, Publications in Anthropology, 8: xii + 1-251.

Carlson, G. F., J. R. Bozell, T. L. Steinacher, M. B. Lovvorn, and G. W. Gill. 1999. The Sidney burial: a Middle Plains Archaic mortuary site from western Nebraska. Plains Anthropologist, 44: 105-119.

Carter, F. L. 1939. A study in jackrabbit shifts in range in western Kansas. Transactions of the Kansas Academy of Science, 42: 431-435.

Champe, J. L. 1936. The Sweetwater Culture complex. Pp. 249-299, *in* Chapters in Nebraska archaeology, ed. E. H. Bell. Lincoln: University of Nebraska, 427 pp.

Champe, J. L. 1946. Ash Hollow Cave: A study of stratigraphic sequence in the central Great Plains. University of Nebraska Studies (new series), 1: ix + 1-130.

Champe, J. L. 1949. White Cat Village. American Antiquity, 14: 285-292.

Choate, J. R., and H. H. Genoways. 1967. Notes on some mammals from Nebraska. Transactions of the Kansas Academy of Science, 69: 238-341.

Choate, J. R., and T. W. Haner. 1992. Probable distribution of the woodchuck in north central Kansas. Prairie Naturalist, 24: 65-66.

Choate, J. R., and K. M. Reed. 1986. Historical biogeography of the woodchuck in Kansas. Prairie Naturalist, 18: 37-42.

Clark, C. P. 1993. Archeological survey and testing at Agate Fossil Beds National Monument, Sioux County, Nebraska. National Park Service, Midwest Archeological Center, Technical Report, 22: vi + 1-82.

Clark, C. P. 1994a. Archeological investigations on Marsland Road, Agate Fossil Beds National Monument, Sioux County, Nebraska. National Park Service, Midwest Archeological Center, Technical Report, 31: v +1-63.

Clark, C. P. 1994b. Archeological survey of Scotts Bluff National Monument, Scotts Bluff County, Nebraska. National Park Service, Midwest Archeological Center, Occasional Studies in Anthropology, 31: vi + 1-66.

Cockrum, E. L. 1948. The distribution of the hispid cotton rat in Kansas. Transactions of the Kansas Academy of Science, 51: 306-312.

Cooper, P. 1936. Archaeology of certain sites in Cedar County, Nebraska. Pp. 15-145, *in* Chapters in Nebraska archaeology, ed. E. H. Bell. Lincoln: University of Nebraska, 427 pp.

Cooper, P. L. 1939. Report of explorations of 1938. Nebraska History, 20: 95-152.

Corner, R. G. 1977. A Late Pleistocene-Holocene vertebrate fauna from Red Willow County, Nebraska. Transactions of the Nebraska Academy of Sciences, 4: 77-93.

Cullingham, C. I. 2008. Genetic structure of raccoons in eastern North America based on mtDNA: implications for subspecies designation and rabies disease dynamics. Canadian Journal of Zoology, 86: 947-958.

Cumming, R. B., Jr., and R. L. Stephenson. 1953. Appraisal of the archeological and paleontological resources of the lower Platte Basin, Nebraska: supplement. Missouri River Basin Project, Smithsonian Institution, Washington, DC, March, 10 pp.

Davis, E. M. 1962. Archeology of the Lime Creek Site in southwestern Nebraska. Special Publication, University of Nebraska State Museum, 3: 1-106.

Davis, M. B., and R. G. Shaw. 2001. Range shifts and adaptive responses to Quaternary climate change. Science, 292: 673-679.

Dunlevy, M. L. 1936. A comparison of the cultural manifestations of the Burkett (Nance County) and the Gray-Wolfe (Colfax County) sites. Pp. 151-247, *in* Chapters in Nebraska archaeology, ed. E. H. Bell. Lincoln: University of Nebraska, 427 pp.

Ewing, S. 2000. Nebraska site DO2: a faunal analysis of birds, fish, rodents and small mammals. Report on file at the Nebraska State Historical Society, Lincoln, 13 pp.

Ezzo, J. A., and M. C. Stiner. 2000. A later Archaic period dog burial from the Tucson Basin, Arizona. Kiva, 66: 291-305.

Falk, C. R. 1969a. Mowry Bluff artifacts: bone, antler, and shell artifacts. Pp. 39-44, *in* Two house sites in the central plains: An experiment in archaeology, ed. W. R. Wood. Plains Anthropologist Memoir, 6: ix + 1-132.

Falk, C. R. 1969b. Mowry Bluff artifacts: faunal remains. Pp. 44-51, *in* Two house sites in the central plains: an experiment in archaeology, ed. W. R. Wood. Plains Anthropologist Memoir, 6: ix + 1-132.

Falk, C. R., and R. E. Pepperl. 1981. Cultural and paleontological resource investigations within the Calamus and Davis Creek Reservoir Areas, Nebraska. Division of Archeological Research, Department of Anthropology, University of Nebraska-Lincoln, Technical Report No. 80-04, 2 vols.

Farney, J. P. 1975. Natural history and northward dispersal of the hispid cotton rat in Nebraska. Platte Valley Review, 3: 11-16.

FAUNMAP Working Group. 1996. Spatial response of mammals to late Quaternary environmental fluctuations. Science, 272: 1601-1608.

Frantz, W. 1963. Four Aksarben sites in Dakota County, Nebraska. Unpublished MA thesis, University of Nebraska, Lincoln, xi + 224 pp.

Frantz, W. 1965. Results of the 1963-1964 field seasons of the Nebraska Archeological Highway Salvage Program. Report on file at the Nebraska State Historical Society, Lincoln, 76 pp.

Freed, M. L. 1954. The Lynch Site, 25BD1. Unpublished M.A. thesis, University of Nebraska, Lincoln, viii + 167 pp.

Freeman, P. W., and H. H. Genoways. 1998. Recent northern records of the nine-banded armadillo (Dasypodidae) in Nebraska. Southwestern Naturalist, 43: 491-495.

Fricke, K. A., M. A. Cover, S. E. Hygnstrom, S. R. Groepper, H. H. Genoways, K. M. Hams, and K. C. VerCauteren. 2008. Historic and recent distribution of elk in Nebraska. Great Plains Research, 18: 189-204.

Galindo, J. L. 1997. Bone element description. Pp. 120-126, *in* Agate Fossil Beds prehistoric archaeological landscapes, 1994-1995, eds. L. Wandsnider and G. H. MacDonell. Lincoln, NE: Department of Anthropology, University of Nebraska-Lincoln, Report for National Park Service, Midwest Archeological Center, viii + 140 pp.

Gant, R. D. 1966. A preliminary appraisal of the archeological resources of the Highway Salvage Program 1966. Report on file at the Nebraska State Historical Society, Lincoln, 113 pp.

Garrett, J. W. 1965. The Birdwood Culture of the west-central plains. American Antiquity, 31: 74-80.

Genoways, H. H., and J. R. Choate. 1972. A multivariate analysis of systematic relationships among populations of the short-tailed shrew (genus *Blarina*) in Nebraska. Systematic Zoology, 21:106-116.

Genoways, H. H., M. J. Hamilton, D. M. Bell, R. R. Chambers, and R. D. Bradley. 2008a. Hybrid zones, genetic isolation, and systematics of pocket gophers (genus *Geomys*) in Nebraska. Journal of Mammalogy, 89: 826-836.

Genoways, H. H., J. D. Hoffman, P. W. Freeman, K. Geluso, R. A. Benedict, and J. J. Huebschman. 2008b. Mammals of Nebraska: checklist, key, and bibliography. Bulletin of the University of Nebraska State Museum, 23: 1-92, 1 October.

Genoways, H. H., and B. C. Ratcliffe. 2008. Engineer Cantonment, Missouri Territory, 1819-1820: America's first biodiversity inventory. Great Plains Research, 18: 3-31.

Genoways, H. H., and D. A. Schlitter. 1967. Northward dispersal of the hispid cotton rat in Nebraska and Missouri. Transactions of the Kansas Academy of Science, 69: 356-357.

George, S. B., J. R. Choate, and H. H. Genoways. 1981. Distribution and taxonomic status of *Blarina hylophaga* Elliot (Insectivora: Soricidae). Annals of Carnegie Museum, 50:493-513.

Giessen, M. 1988. Processual analysis of the bone, antler and shell artifact assemblage from house 2 at Annie's Site. Pp. 133-156, *in* St. Helena archaeology: new data, fresh interpretations, ed. D. J. Blakeslee. Reprints in Anthropology, 39: vi + 1-193.

Gilder, R. F. 1907. Archeology of the Ponca Creek District, eastern Nebraska. American Anthropologist, 9: 702-719.

Gilder, R. F. 1908. Recent excavation at Long's Hill, Nebraska. American Anthropologist, 10: 60-73.

Gilder, R. F. 1909. Excavation of earth-lodge ruins in eastern Nebraska. American Anthropologist, 11: 56-84.

Gillaspie, E. 1955. A report on the Sheep Mountain Site, 25BN1, Banner Co., Nebraska. Report on file at the Nebraska State Historical Society, Lincoln, 24 pp.

Goedert, A. J. 1995. Summary of faunal remains. Pp. 166-171, *in* The Stabaco Site: a mid-eighteenth century Skidi Pawnee town on the Loup River, eds. S. R. Holen and J. K. Peterson. Nebraska Archeological Survey, University of Nebraska State Museum, Technical Report, 95-01: iv + 1-221.

Gradwohl, D. M. 1969. Prehistoric villages in eastern Nebraska. Nebraska State Historical Society, Publications in Anthropology, 4: xii + 1-203.

Graham, R. W., H. A. Semken, Jr., and M. A. Graham. 1987. Late Quaternary mammalian biogeography of the Great Plains and prairies. Scientific Papers, Illinois State Museum, 22: xiv + 1-491.

Grange, R. T., Jr. 1980. Salvage archeology in the Red Willow Reservoir, Nebraska. Nebraska State Historical Society, Publications in Anthropology, 9: xi + 1-236.

Griffin, K. 1991. Test excavation at site 25SF50, Scotts Bluff National Monument, Nebraska. National Park Service, Midwest Archeological Center, Technical Report, 5: vi + 1-83.

Groves, C., and P. Grubb. 2011. Ungulate taxonomy. Baltimore: Johns Hopkins University Press, ix + 317 pp.

Gunnerson, D. A. n.d. [1956]. The Stanton Site, 25ST1. Unpublished masters thesis, University of Nebraska, Lincoln, pages unnumbered, on file Nebraska State Historical Society, Lincoln.

Gunnerson, J. H. 1960. An introduction to Plains Apache archeology—The Dismal River Aspect. Smithsonian Institution, Bureau of American Ethnology Bulletin, 173: 131-260.

Gunnerson, J. H., and D. A. Gunnerson. 1952. Appraisal of the archeological and paleontological resources of the lower Platte Basin, Nebraska: supplement. Missouri River Basin Project, Smithsonian Institution, Washington, DC, September, 27 pp.

Haas, D. R. 1983. Walker Gilmore: a stratified Woodland period occupation in eastern Nebraska, a report of the 1968 excavations. Division of Archeological Research, Department of Anthropology, University of Nebraska-Lincoln, Notebook, 6: x +1-260.

Hafner, J. C., and M. S. Hafner. 1983. Evolutionary relationships of heteromyid rodents. Great Basin Naturalist Memoirs 7:3-29.

Hall, E. R. 1981. The mammals of North America. New York: John Wiley & Sons, 1: xv + 1-600 + *90*; 2: vi + 601-1181 + *90*.

Helgren, K. M., F. R. Cole, L. E. Helgren, and D. E. Wilson. 2009. Generic revision in the Holoarctic ground squirrel genus *Spermophilus*. Journal of Mammalogy, 90: 270-305.

Heaney, L. R. and R. M. Timm. 1983. Relationships of pocket gophers of the genus *Geomys* from the central and Great Plains. Miscellaneous Publication of the Museum of Natural History, University of Kansas, 74:1-59.

Heaney, L. R., and R. M. Timm. 1985. Morphology, genetics, and ecology of pocket gophers (genus *Geomys*) in a narrow hybrid zone. Biological Journal of the Linnean Society, 25: 301-317.

Henning, D. R. 1977. Archaeological investigations in the proposed midstate irrigation project. Midstate Archaeological Survey, National Park Service, 126 pp.

Hill, A. T. 1927. Mr. A. T. Hill's own story. Nebraska History Magazine, 10: 162-167.

Hill, A. T. 1932. Ruins of a prehistoric house in Howard County, Nebraska. Nebraska History Magazine, 13: 172-175.

Hill, A. T., and P. Cooper. 1936a. The Schrader Site: prehistoric village in Lancaster County, Nebraska. Nebraska History Magazine, 17: 222-252.

Hill, A. T., and P. Cooper. 1936b. The Champe Site: excavation of a prehistoric house in Douglas County, Nebraska. Nebraska History Magazine, 17: 253-270.

Hill, A. T., and P. Cooper. 1936c. Fremont 1: prehistoric village site in Sarpy County, Nebraska. Nebraska History Magazine, 17: 271-292.

Hill, A. T., and P. Cooper. 1937. The archeological campaign of 1937 by the Nebraska State Historical Society. Nebraska History Magazine, 18: 243-359.

Hill, A. T., and M. Kivett. 1940. Woodland-like manifestations in Nebraska. Nebraska History, 21: 143-243.

Hill, A. T., and G. Metcalf. 1942. A site of the Dismal River Aspect in Chase County, Nebraska. Nebraska History, 22: 158-226.

Hill, A. T., and W. R. Wedel. 1936. Excavations at the Leary Indian village and burial site, Richardson County, Nebraska. Nebraska History Magazine, 17: 2-73.

Hill, M. A. 1988. The bone tool assemblage of the St. Helena phase. Pp. 99-132, *in* St. Helena archaeology: new data, fresh interpretations, ed. D. J. Blakeslee. Reprints in Anthropology, 39: vi + 1-193.

Hill, M. E., Jr. 2008. New investigations of the Code-age Finley and Scottsbluff bison bone beds. Current Research in the Pleistocene, 25: 90-93.

Hill, M. G. 2005. Late Paleoindian (Allen Frederick complex) subsistence activities at the Clary Ranch Site, Ash Hollow, Garden County, Nebraska. Plains Anthropologist, 50: 249-263.

Hill, M. G., D. W. May, D. J. Rapson, A. R. Boehn, and E. Otárola-Castillo. 2008. Faunal exploitation by early Holocene hunter/gatherers on the Great Plains of North America: evidence from the Clary Ranch sites. Quaternary International, 191: 115-130.

Hoak, J. H., J. L. Weaver, and T. W. Clark. 1982. Wolverines in western Wyoming. Northwest Science, 56: 159-161.

Hoffman, J. D., and H. H. Genoways. 2005. Recent records of formerly extirpated carnivores in Nebraska. Prairie Naturalist, 37: 225-244.

Hoffman, J. D., H. H. Genoways, and J. R. Choate. 2006. Long-distance dispersal and population trends of moose in the central United States. Alces, 42: 115-131.

Hoffman, J. D., H. H. Genoways, and R. R. Jones. 2011. Historical biogeography of Nebraska pronghorns (*Antilocapra americana*). Great Plains Research, 21: 153-173.

Hoffman, J. D., S. Wilson, and H. H. Genoways. 2009. Recent occurrence of an American black bear in Nebraska. Ursus, 20: 69-72.

Hoffmann, R. S., and J. K. Jones, Jr. 1970. Influence of the late-glacial and post-glacial events on the distribution of Recent mammals on the Northern Great Plains. Pp. 355-394, *in* Pleistocene and Recent environments of the Central Great Plains, eds. W. Dort, Jr., and J. K. Jones, Jr. Special Publication, Department of Geology, University of Kansas, 3: 1-433.

Holder, P., and J. Wike. 1949. The Frontier Culture Complex, a preliminary report on a prehistoric hunters' camp in southwestern Nebraska. American Antiquity, 14: 260-266.

Holen, S. R. 1995. The Stabaco Site: An introduction. Pp. 1-13, *in* The Stabaco Site: a mid-eighteenth century Skidi Pawnee town on the Loup River, eds. S. R. Holen and J. K. Peterson. Nebraska Archeological Survey, University of Nebraska State Museum, Technical Report, 95-01: iv + 1-221.

Holen, S. R., and D. W. May. 1989. Report on preliminary investigations at 25FT177: a probable late Pleistocene man/mammoth association. Report to the Bureau of Reclamation, Grand Island Office, Grand Island, NE, 17 pp.

Holen, S. R., D. W. May, and P. M. Prettyman. 1996. An archaeological and geomorphic survey of Paleoindian sites at Harlan County Lake, Harlan County, Nebraska. University of Nebraska State Museum, Nebraska Archaeological Survey, 96-01: iv +1-73.

Holen, S. R., and J. K. Peterson. 1995. Conclusion. Pp. 219-221, *in* The Stabaco Site: a mid-eighteenth century Skidi Pawnee town on the Loup River, eds. S. R. Holen and J. K. Peterson. Nebraska Archeological Survey, University of Nebraska State Museum, Technical Report, 95-01: iv + 1-221.

Holen, S. R., and D. R. Watson. 1999. Phase I survey and Phase II testing of archaeological sites at the Arcadia diversion dam lands on the Middle Loup River, Custer County, Nebraska. Nebraska Archaeological Survey, University of Nebraska State Museum, Technical Report, 99-01: vi + 1-113.

Holen, S. R., and D. R. Watson. 2000. An archaeological survey of the lower South Platte River and Pumpkin Creek drainage. Nebraska Archaeological Survey, University of Nebraska State Museum, Technical Report, 2000-02: iv + 1-123.

Holen, S. R., D. R. Watson, and J. K. Peterson. 2001. Methodology and site description. Pp. 1-44, *in* Central Plains Tradition at the Palmer Locality: features and artifacts from 25NC29, eds. J. K. Peterson, S. R. Holen, and D. R. Watson. Nebraska Archaeological Survey, University of Nebraska State Museum, 2001-01: xiii + 1-229.

Howard, J. H., and R. D. Gant. 1966. Archeological salvage investigations in the Gavin's Point Reservoir area, Lewis and Clark Lake, Nebraska and South Dakota, 1963 and 1964. South Dakota Museum, University of South Dakota, Archeological Studies Circular, 11: 1-70.

Huebschman, J. J. 2007. Distribution, abundance, and habitat associations of Franklin's ground squirrel (*Spermophilus franklinii* (Sabine 1822). Bulletin of the Illinois Natural History Survey, 38(1): ii + 1-57.

Hudson, J. 2007. Faunal evidence for subsistence and settlement patterns at the Allen Site. Pp. 194-226, *in* The Allen Site: a Paleoindian camp in southwestern Nebraska, ed. D. B. Bamforth. Albuquerque: University of New Mexico Press, xvi + 284 pp.

Jacobson, J. A. 2003. Identification of mule deer (*Odocoileus hemionus*) and white-tailed deer (*Odocoileus virginianus*) postcranial remains as a means of determining human subsistence strategies. Plains Anthropologist, 48: 287-297.

Jensen, R. J. 1973. Archeological salvage and survey in Nebraska: Part II: A preliminary report of the Point of Rocks Archeological Survey. Nebraska State Historical Society, Publications in Anthropology, 5: 159-240.

Jones, C. A., J. R. Choate, and H. H. Genoways. 1984. Phylogeny and paleobiogeography of short-tailed shrews (genus *Blarina*). Pp. 56-148, *in* Contributions in Quaternary vertebrate paleontology: a volume in memorial to John E. Guilday, eds. H. H. Genoways and M. R. Dawson. Special Publication, Carnegie Museum of Natural History, 8:1-538.

Jones, J. K., Jr. 1960. The hispid cotton rat in Nebraska. Journal of Mammalogy, 41: 132.

Jones, J. K., Jr. 1964. Distribution and taxonomy of mammals of Nebraska. Museum of Natural History, University of Kansas Publications, 16: 1-356.

Kaldahl, E. 1993. The Nebraska Phase: an island of stability in a region of relative change. Unpublished undergraduate honors thesis, Department of Anthropology, University of Nebraska, Lincoln, v + 112 pp.

Kaufman, D. W., R. A. Kaufman, and G. A. Kaufman. 2016. First record of the woodchuck in Osborne County, Kansas. Transactions of the Kansas Academy of Science, 119: 336-338.

Kay, M. 1975. Archeological surveys in Scotts Bluff and Agate Fossil Beds National Monuments, Nebraska. Division of Archeological Research, Department of Anthropology, University of Nebraska-Lincoln, Technical Report, 76-13: iv + 1-54.

Kivett, M. F. 1947. Preliminary appraisal of the archeological and paleontological resources of certain proposed reservoir areas in the lower Platte sub-basin Nebraska. River Basin Surveys, Smithsonian Institution, September, 39 pp.

Kivett, M. F. 1951. Archeological investigations Swanson Lake Project, Hitchcock County, southwestern Nebraska in 1950. Interagency Archeological Services, Denver, CO, v + 73 pp.

Kivett, M. F. 1952. Woodland sites in Nebraska. Nebraska State Historical Society, Publications in Anthropology, 1: ix + 1-102.

Kivett, M. F. 1961. Preliminary appraisal of the archaeological resources of the Red Willow Reservoir, Hayes, Frontier, and Red Willow counties, Nebraska. A Project of the Inter-agency Archeological and Paleontological Salvage Program, Nebraska State Historical Society, 27 pp.

Kivett, M. F. 1959. Logan Creek complex, site 25BT3. Nebraska State Historical Society, November, 5 pp.

Kivett, M. F., and G. S. Metcalf. 1997. The prehistoric people of the Medicine Creek Reservoir, Frontier County, Nebraska: an experiment in mechanized archeology (1946-1948). Plains Anthropologist, 42: i-v + 1-218. [preface by D. M. Gradwohl]

Knudson, R. 2002. Medicine Creek is a Paleoindian cultural ecotone: The Red Smoke assemblage. Pp. 84-141, *in* Medicine Creek: Seventy years of archaeological investigations, ed. D. C. Roper. Tuscaloosa: University of Alabama Press, xix + 248 pp.

Knudson, R. 2013. The Scottsbluff Bison Quarry site: its place in the Cody complex. Pp. 290-314, *in* Paleoindian lifeways of the Cody complex, eds. E. J. Knell and M. P. Muñoz. Salt Lake City: University of Utah Press, xii + 340 pp.

Koch, A. 1995. The McIntosh fauna: late prehistoric exploitation of lake and prairie habitats in the Nebraska Sand Hills. Plains Anthropologist, 40: 39-60.

Koch, A. 2000. High plains archeology. Explore Nebraska Archeology, 5: 1-20.

Koch, A. 2002. Archeology of the Beaver Creek Site, Boone County, Nebraska. Nebraska State Historical Society, Report to the Nebraska Department of Roads, x + 161 pp.

Koch, A. 2004. McIntosh: a late prehistoric occupation in the Nebraska Sand Hills. Central Plains Archeology, 10 (1): xii + 1-167.

Koch, A., and J. Miller. 1996. Georcheological investigations at the Lyman Site (25SF53) and other cultural resources related to Table Mountain Quarry near the Nebraska/Wyoming border. Nebraska State Historical Society, Report for the Nebraska Department of Roads, ix + 153 pp.

Koch, A., and T. Nelson. 2002. Material recovered. Pp. 19-108, *in* Archeology of the Beaver Creek Site, Boone County, Nebraska, ed. A. Koch. Nebraska State Historical Society, Report to the Nebraska Department of Roads, x + 161 pp.

Koepfli, K.-P., and R. K. Wayne. 1998. Phylogenetic relationships of otters (Carnivora: Mustelidae) based on mitochondrial cytochrome b sequences. Journal of Zoology, 246: 401-416.

Landholt, L. M., and H. H. Genoways. 2000. Population trends in furbearers in Nebraska. Transactions of the Nebraska Academy of Sciences, 26: 97-110.

Lamb, G. F. 1932. Earth lodge ruins of Indian habitation on Rose Creek in Thayer County. Nebraska History Magazine, 13: 169-172.

Lamb, G. F. 1939a. Bakenhus Site. Report on file at Nebraska State Historical Society, Lincoln, 40 pp.

Lamb, G. F. 1939b. Bakenhus ossuary. Report on file at Nebraska State Historical Society, Lincoln, 19 pp.

Linzey, A. V. 1983. *Synaptomys cooperi*. Mammalian Species, 210: 1-5.

Ludt, C. J., W. Schroeder, O. Rottmann, and R. Kuehn. 2004. Mitochondrial DNA phylogeography of red deer (*Cervus elaphus*). Molecular Phylogenetics and Evolution, 31: 1064-1083.

Ludwickson, J. 1978a. Central Plains Tradition settlements in the Loup River Basin: the Loup River Phase. Pp. 94-108, *in* The Central Plains Tradition: Internal development and external relationships, ed. D. J. Blakeslee. Office of the State Archaeologist, University of Iowa, Iowa City, Report 11: 1-166.

Ludwickson, J. 1978b. Artifact descriptions Midstate Survey. Midstate Archaeological Survey, National Park Service, iii + 59 pp.

Ludwickson, J., D. Blakeslee, and J. O'Shea. 1981. Missouri National Recreational River: Native American cultural resources. Report for Heritage Conservation and Recreation Service, Interagency Archeological Services, Denver, CO, iv + 292 pp.

Ludwickson, J., J. R. Bozell, D. Bringelson, and A. Koch. 1994. Material recovered. Pp. 108-140, *in* Nebraska phase archeology in the South Bend locality, eds. J. R. Bozell and J. Ludwickson. Nebraska State Historical Society, Report for Nebraska Department of Roads, xiii + 236 pp.

Lyons, S. K. 2003. A quantitative assessment of the range shifts of Pleistocene mammals. Journal of Mammalogy, 84: 385-402.

Lyons, S. K., P. J. Wagner, and K. Dzikiewicz. 2010. Ecological correlates of range shifts of Late Pleistocene mammals. Philosophical Transactions: Biological Sciences, 365: 3681-3693.

Manz, K., and D. J. Blakeslee. 1988. An analysis of the faunal remains from house 2, 25DX30. Pp. 157-171, *in* St. Helena archaeology: new data, fresh interpretations, ed. D. J. Blakeslee. Reprints in Anthropology, 39: vi + 1-193.

Mayewski, P. A., E. E. Rohling, J. C. Stager, W. Karlen, K. A. Maasch, L. D. Meeker, E. A. Mayerson, F. Gasse, S. van Kreveld, K. Holmgren, J. Lee-Thorp, G. Rosqvist, F. Rack, M.

Staubwasser, R. R. Schneider, and E. J. Steig. 2004. Holocene climate variability. Quaternary Research, 62: 243-255.

Meadow, B., and J. K. Peterson. 2001. Unmodified and modified faunal remains. Pp. 154-177, *in* Central Plains Tradition at the Palmer Locality: features and artifacts from 25NC29, eds. J. K. Peterson, S. R. Holen, and D. R. Watson. Nebraska Archaeological Survey, University of Nebraska State Museum, 2001-01: xiii + 1-229.

Meston, L. K. 1976. Archaeological investigations in the White River region, northwest Nebraska. Division of Archeological Research, Department of Anthropology, University of Nebraska-Lincoln, Technical Report No. 76-12: ii +1-58.

Metcalf, G. 1941a. The Hill-Rupp (or Monroe) Site. Report on file at Nebraska State Historical Society, Lincoln, 34 pp.

Metcalf, G. 1941b. The Hill Farm or Pike Pawnee Site (WT1). Report on file at Nebraska State Historical Society, Lincoln.

Mick, L. S. 1983. An ecological evaluation of faunal diversity in the Central Plains tradition. Unpublished M.A. thesis, Department of Anthropology, University of Nebraska-Lincoln, vi + 263 pp.

Moore, K. R. 1988. A cultural resources sample survey in the Harlan County Lake project lands west of U. S. Highway 183, Harlan County, Nebraska. American Resources Group, Ltd., Carbondale, IL, Cultural Resources Management Report, 70: I + 1-146.

Morey, D. F. 1994. The early evolution of the domestic dog. American Scientist, 82: 336-347.

Moulton, M. P., J. R. Choate, and S. J. Bissell. 1979. Sympatry of pocket gophers on Mesa de Maya, Colorado. Transactions of the Kansas Academy of Science, 82: 194-195.

Nadler, C. F., R. S. Hoffmann, and K. R. Greer. 1971. Chromosomal divergence during evolution of ground squirrel populations (Rodentia: *Spermophilus*). Systematic Zoology, 20: 298-305.

Nelson, T. (ed.). 2006. Archeological investigations at the Rulo Southeast sites: late pre-contact occupation along the Missouri River, Richardson County, Nebraska. Nebraska State Historical Society, Highway Archeology Program for the Nebraska Department of Roads [Project STPE-3625(3)], ix + 220 pp.

Nepstad-Thornberry, C., L. S. Cummings, and K. Puseman. 2002. A model for the Upper Republican subsistence and nutrition in the Medicine Creek locality: a new look at extant data. Pp. 197-211, *in* Medicine Creek: seventy years of archaeological investigations, ed. D. C. Roper. Tuscaloosa: University of Alabama Press, xix + 248 pp.

Oothoudt, J. W. 1976. The "Gering" (25SF10) and "Dry Lake" (25MP2) burials: chronologic and cultural implications. Unpublished M.A. thesis, University of Nebraska, Lincoln, vii + 177 pp.

Ore, H. K. 1981. The White Site. Report on file at the Nebraska State Historical Society, 20 pp.

Osborn, A. J. 1979. Cultural resources inventory and assessment for select areas within the Ft. Niobrara National Wildlife Refuge, Valentine, Nebraska: a final report. Division of Archeological Research, Department of Anthropology, University of Nebraska-Lincoln, Technical Report No. 79-07, 241 pp.

O'Shea, J. M., and J. Ludwickson. 1992, Archaeology and ethnohistory of the Omaha Indians: the Big Village site. Lincoln: University of Nebraska Press, xviii + 374 pp.

Pepperl, R. E., and C. R. Falk. 1983. Native American archeological resources of the central Niobrara River valley: the Norden study, north central Nebraska. Division of Archeological Research, Department of Anthropology, University of Nebraska-Lincoln, Technical Report No. 82-10: xv + 1-51, A1-A389, B1-B99, C1-C27.

Peterson, J. K., and S. R. Holen (eds.). 1996. The Tahaksu Site (25MK15), Merrick County, Nebraska. Nebraska Archaeological Survey, University of Nebraska State Museum, Technical Report, 96-03: vi + 1-162.

Peterson, J. K., and S. R. Holen. 2001. Site interpretation and conclusions. Pp. 178-182, *in* Central Plains Tradition at the Palmer Locality: features and artifacts from 25NC29, eds. J. K. Peterson, S. R. Holen, and D. R. Watson. Nebraska Archaeological Survey, University of Nebraska State Museum, 2001-01: xiii + 1-229.

Philips, R. E. 1954. Scheivalbein Site, Elm Creek 2, Gosper County. Report on file at Nebraska State Historical Society, Lincoln, 8 pp.

Philips, R. E. 1956a. Helms Site, Muddy Creek MD 4, Gosper County. Report on file at Nebraska State Historical Society, Lincoln, 1 p.

Philips, R. E. 1956b. Gross Site, Muddy Creek MD-9, Gosper County. Report on file at Nebraska State Historical Society, Lincoln, 5 pp.

Philips, R. E. 1957a. Mousel Site, Turkey Creek I, Gosper Co. Report on file at Nebraska State Historical Society, Lincoln, 11 pp.

Philips, R. E. 1957b. Golter Site, Elm Creek 1, Gosper County. Report on file at Nebraska State Historical Society, Lincoln, 4 pp.

Philips, R. E. 1957c. Graf Site, Muddy Creek MD 5, Gosper County. Report on file at Nebraska State Historical Society, Lincoln, 2 pp.

Philips, R. E. 1958. Reuben Lucas Site, Muddy Creek-8. Report on file at Nebraska State Historical Society, Lincoln, 3 pp.

Philips, R. E. 1959. Lowell Dawson Site, Turkey Creek 9 – Gosper County. Report on file at Nebraska State Historical Society, Lincoln, 5 pp.

Philips, R. E. 1960a. The Dickman Ossuary, a prehistoric burial site in Custer County, Nebr. Report on file at Nebraska State Historical Society, Lincoln, 22 pp.

Philips, R. E. 1960b. Stillman Site, Muddy Creek-3. Report on file at Nebraska State Historical Society, Lincoln, 4 pp.

Philips, R. E. 1960c. Greisfeller Site, Muddy Creek 7. Report on file at Nebraska State Historical Society, Lincoln, 3 pp.

Philips, R. E. 1960d. Dawson Site, Turkey Creek T-6. Report on file at Nebraska State Historical Society, Lincoln, 7 pp.

Polziehn, R. O., and C. Strobeck. 2002. A phylogenetic comparison of red deer and wapiti using mitochondrial DNA. Molecular Phylogenetics and Evolution, 22: 342-356.

Price, R. S. 1956. Early ceramic period sites in northeastern Nebraska. Unpublished M.A. thesis, University of Nebraska, Lincoln, ix + 218 pp.

Reed, K. M., and J. R. Choate. 1986. Geographic variation in the plains pocket mouse (*Perognathus flavescens)* on the Great Plains. The Texas Journal of Science, 38: 227-240.

Reichman, O. J., and R. J. Baker. 1972. Distribution and movements of two species of pocket gophers (Geomyidae) in an area of sympatry in the Davis Mountains, Texas. Journal of Mammalogy, 53: 21-33.

Reitz, E., and N. Honerkamp. 1983. British colonial subsistence strategy on the southeastern coastal plain. Historical Archaeology, 17: 4-26.

Renaud, E. B. 1933. Archaeological survey of western Nebraska. Department of Anthropology, University of Denver, 59 pp.

Roehrs, Z. P., R. A. Benedict, T. E. Labedz, and H. H. Genoways. 2021. Observations on the distribution and status of selected Nebraska mammals. Transactions of the Nebraska Academy of Sciences, 41: 1-28.

Roehrs, Z. P., and H. H. Genoways. 2004. Historical biogeography of the woodchuck (*Marmota monax bunkeri*) in Nebraska and northern Kansas. Western North American Naturalist, 64: 396-402.

Roper, D. C. 1989. Protohistoric Pawnee hunting in the Nebraska Sand Hills: archeological investigations at two sites in the Calamus Reservoir. Report to the Bureau of Reclamation, Great Plains Region, Billings, MT, iii + 328 pp.

Roper, D. C. 1994. The material culture of 25DS21, a Lower Loup hunting camp in the Platte River valley. Central Plains Archaeology, 4: 55-95.

Roper, D. C. 1998. The Schudel Site complex: Early Archaic occupation in the North Loup River Valley. Central Plains Archaeology, 6: 1-34.

Rusco, M. K. 1960. The White Rock Aspect. University of Nebraska, Laboratory of Anthropology, Note Book, 4: 1-97.

Satorius-Fox, M. R. 1982. Paleoecological analysis of micromammals from the Schmidt Site, a Central Plains Tradition village in Howard County, Nebraska. Division of Archeological Research, Department of Anthropology, University of Nebraska-Lincoln, Technical Report No. 82-13: vi + 1-87.

Schultz, C. B., and L. Eiseley. 1935. Paleontological evidence for the antiquity of the Scottsbluff Bison Quarry and its associated artifacts. American Anthropologist, new series, 37: 306-319.

Semken, H. A., Jr. 1983. Holocene mammalian biogeography and climatic change in the eastern and central United States. Pp. 182-207, *in* Late-Quaternary environments of the United States. Vol. 2. The Holocene, ed. H. E. Wright. Minneapolis: University of Minnesota Press, xvil + 277 pp.

Silver, J. 1927. The introduction and spread of house rats in the United States. Journal of Mammalogy, 8: 58-60.

Snyder, L. M., and J. R. Bozell. 1983. Identification and analysis of vertebrate faunal remains recovered from the Logan Creek Site (25BT3) Burt County, Nebraska. Division of Archeological Research, Department of Anthropology, University of Nebraska-Lincoln, Technical Report No. 83-03: vi + 1-113 .

Sounders, P. 1994. Sympatry-mapping studies of small mammal remains recovered from six prehistoric Central Plains sites. Central Plains Archeology, 4: 43-54.

Statham, M. J., J. Murdock, J. Janecka, K. B. Aubry, C. J. Edwards, C. D. Soulsbury, O. Berry, Z. Wang, D. Harrison, M. Pearch, L. Tomsett, J. Chupasko, and B. N. Sacks. 2014. Range-wide multilocus phylogeography of the red fox reveals ancient continental divergence, minimal genomic exchange and distinct demographic histories. Molecular Ecology, 23: 4813-4830.

Statham, M. J., B. N. Sacks, K. B. Aubry, J. D. Perrine, and S. M. Wisely. 2012. The origin of recently established red fox populations in the United States: translocations or natural range expansion? Journal of Mammalogy, 93: 52-65.

Steinacher, T. L., and G. F. Carlson. 1984. Nebraska Highway Archeological and Historical Salvage Investigations, 1969-1975. Nebraska State Historical Society, Publications in Anthropology, 10: xiv + 1-191.

Sterns, F. H. 1915. A stratification of cultures in eastern Nebraska. American Anthropologist, 17: 121-127.

Stewart, J. D. 1987. Paleontology and paleoecology of the North Cove site, 25HN164. Pp. 298-335, *in* Prehistoric and historic cultural resources of selected sites at Harlan County Lake, Harlan County, Nebraska: test excavations and determination of significance for 28 sites, eds. M. J. Adair and K. L. Brown. Kaw Valley Engineering and Development, Inc., Junction City, KS, xiii + 594 pp.

Strong, W. D. 1935. An introduction to Nebraska archeology. Smithsonian Miscellaneous Collections, 93(10): vii + 1-323.

Sudman, P. D., J. R. Choate, and E. G. Zimmerman. 1987. Taxonomy of chromosomal races of *Geomys bursarius lutescens* Merriam. Journal of Mammalogy, 68: 526-543.

Sudman, P. D., J. K. Wickcliffe, P. Horner, M. J. Smolen, J. W. Bickham, R. D. Bradley. 2006. Molecular systematics of pocket gophers of the genus *Geomys*. Journal of Mammalogy, 87: 668-676.

Teter, D. 1975. Archaeological survey of the Boyd County pumped storage project. Division of Archeological Research, Department of Anthropology, University of Nebraska-Lincoln, Technical Report No. 75-03: 1-169.

Than, K. 2011. Oldest domesticated dog in Americas found—was human food. National Geographic New, accessed at http://news.nationalgeographic.com/new/2011/01/110118/, posted 17 January 2011.

Thompson, C. W., R. S. Pfau, J. R. Choate, H. H. Genoways, and E. J. Finck. 2011. Identification and characterization of the contact zone between short-tailed shrews (*Blarina*) in Iowa and Missouri. Canadian Journal of Zoology, 89: 278-288.

Turnmire, K. 1996. Unmodified fauna. Pp. 187-250, *in* Toward a new perspective on Upper Republican life in the Medicine Creek valley, ed. D. C. Roper. Report to the U. S. Department of the Interior, Bureau of Reclamation, Great Plains Region, 389 pp.

Van Zyll de Jong, C. G. 1972. A systematic review of the Nearctic and Neotropical river otters (Genus *Lutra*, Mustelidae, Carnivora). Life Sciences Contributions of the Royal Ontario Museum, 80: 1-104.

Van Zyll de Jong, C. G. 1987. A phylogenetic study of the Lutrinae (Carnivora; Mustelidae) using morphological data. Canadian Journal of Zoology, 65: 2536-2544.

Van Zyll de Jong, C. G. 1991. A brief review of the systematics and a classification of the Lutrinae. Pp. 79-83, *in* Proceedings of the Vth International Otter Colloquium, Habitat 6, eds. C. Reuther and R. Rochert. Hankensbuttel, Federal Republic of Germany.

Viau, A. E., K. Gajewski, M. C. Sawada, and P. Finnes. 2006. Millennial-scale temperature variations in North America during the Holocene. Journal of Geophysical Research, 111, D09102, doi: 10.1029/2005JD006031.

Voorhies, M. R. 1987. Fossil armadillos in Nebraska: the northernmost record. Southwestern Naturalist, 32: 237-243.

Walker, D. N. 2000. Pleistocene and Holocene records of *Antilocapra americana*: a review of the FAUNMAP data. Plains Anthropologist, 45: 13-28.

Watson, D. R. 1996. Chapter 5: Summary of faunal analysis and artifacts. Pp. 128-136, *in* The Tahaksu Site (25MK15), Merrick County, Nebraska, eds. J. K. Peterson and S. R. Holen. Nebraska Archaeological Survey, University of Nebraska State Museum, Technical Report, 96-03: vi + 1-162.

Watson, D. R., J. Dillon, and A. J. Osborn. 2004. An archaeological survey within the Bow Creek drainage, Cedar County, Nebraska. Nebraska Archaeological Survey, University of Nebraska State Museum, Technical Report, 2004-03: unnumbered.

Webber, J. J. Z., and E. J. Reitz. 1999. Animal use on the eighteenth-century frontier Stobo Plantation, South Carolina. Pp. 283-298, *in* Willtown: an archaeological and historical perspective, eds. M. Zierden, S. Linder, and R. Anthony. The Charleston Museum Archaeological Contributions, 27: xiv + 1-369.

Wedel, W. R. 1934. Contributions to the archeology of the Upper Republican Valley, Nebraska. Nebraska History Magazine, 15: 133-209.

Wedel, W. R. 1936. An introduction to Pawnee archeology. Smithsonian Institution, Bulletin of the Bureau of American Ethnology, 112: xi + 1-122.

Wedel, W. R. 1938. The direct-historical approach in Pawnee archeology. Smithsonian Miscellaneous Collections, 97(7): 1-21.

Wedel, W. R. 1986. Central Plains prehistory: Holocene environments and culture change in the Republican River Basin. Lincoln: University of Nebraska Press, xviii + 280 pp.

Wendland, W. M. 1978. Holocene man in North America: the ecological settings and climatic background. Plains Anthropologist, 23: 273-287.

White, T. E. 1951. Appraisal of the archeological and paleontological resources of the Niobrara River Basin, Nebraska. Report on file Midwest Archeological Center, Lincoln, NE, 62 pp.

White, T. E., and P. L. Cooper. 1951. Appraisal of the archeological and paleontological resources of the Niobrara River Basin, Nebraska. Missouri River Basin Survey, Smithsonian Institution, Washington, DC, August, 62 pp.

Widga, C. C. 2004. Early Archaic subsistence in the Central Plains: the Spring Creek (25FT31) fauna. Plains Anthropologist, 49: 25-58.

Widga, C. C. 2006. Bison, bogs, and big bluestem: the subsistence ecology of Middle Holocene hunter-gatherers in the eastern Great Plains. Unpublished Ph.D. dissertation, University of Kansas, Lawrence, xiv + 333 pp.

Williams, S. L., and R. J. Baker. 1976. Vagility and local movements of pocket gophers (Geomyidae: Rodentia). American Midland Naturalist, 96: 303-316.

Wilson, G. M., and J. R. Choate. 1996. Continued westward dispersal of the woodchuck in Kansas. Prairie Naturalist, 28: 21-22.

Wilson, G. M., and J. R. Choate. 1997. Taxonomic status and biogeography of the southern bog lemming, *Synaptomys cooperi*, on the central Great Plains. Journal of Mammalogy, 78: 444-458.

Winfrey, J. V. 1991. Spatial distribution of cultural material and post-depositional disturbances at the Wallace Site (25GO2): a Plains Woodland occupation site in south central Nebraska. Unpublished M.A. thesis, University of Nebraska, Lincoln, ix + 149 pp.

Wisely, S. M., M. J. Statham, and R. C. Fleischer. 2008. Pleistocene refugia and Holocene expansion of the grassland-dependent species, the black-footed ferret (*Mustela nigripes*). Journal of Mammalogy, 89: 87-96.

Witty, T. A. 1962. The Anoka Focus. Unpublished M.A. thesis, University of Nebraska, Lincoln, vi + 192 pp.

Wood, W. R. 1956. The Redbird Focus. Unpublished M.A. thesis, University of Nebraska, Lincoln, xiv + 229 pp.

Wood, W. R. 1965. The Redbird Focus and the problem of Ponca prehistory. Plains Anthropologist, 10: 79-145.

Wood, W. R. 1969a. The Mowry Bluff Site, 25FT35: site description. Pp. 3-6, *in* Two house sites in the central plains: an experiment in archaeology, ed. W. R. Wood. Plains Anthropologist Memoir, 6: ix + 1-132.

Wood, W. R. 1969b. The Mowry Bluff Site, 25FT35: carbon-14 dates. P. 14, *in* Two house sites in the central plains: an experiment in archaeology, ed. W. R. Wood. Plains Anthropologist Memoir, 6: ix + 1-132.

Wood, W. R. 1993. Nánna, the Ponca Fort. 2nd edition. Reprints in Anthropology, J & L Reprint Company, Lincoln, NE, 44: vii + 1-124.

Wright, G. D., K. Geluso, and R. A. Benedict. 2010. Hispid cotton rat (*Sigmodon hispidus*) in Nebraska: distribution, reproduction, and seasonal activity. Western North American Naturalist, 70: 400-406.

Yang, D. Y., J. R. Woiderski, and J. C. Driver. 2005. DNA analysis of archaeological rabbit remains from the American Southwest. Journal of Archaeological Science, 32: 567-578.

Zegers, D. A. 1984. *Spermophilus elegans*. Mammalian Species, 214: 1-7.

Appendix I

Gazetteer of Archeological and Paleontological Sites Where Holocene Mammalian Remains Were Recovered in Nebraska

Antelope Co.

Rosedale Reservoir, 25AP11, Sec. 23, T25N, R8W, west-central Antelope Co., southwest of Clearwater; age—AD 1000-1350 (Gunnerson and Gunnerson, 1952: 21)

25AP37, northwestern edge of Neligh; age—AD 1600-1700 (Buhta et al., 2007: 138)

Banner Co.

Sheep Mountain Site, 25BN1, 5.5 mi. S McGrew [in Scotts Bluff Co.]; age—unknown (Gillaspie, 1955: 1-2)

25BN28, Pumpkin Creek, 7 mi. NE Harrisburg; age—unknown (Steinacher and Carlson, 1984: 2-4)

Boone Co.

Beaver Creek Site, 25BO23, 2.4 km N Loretto; age—AD 1100-1400 (Koch, 2002: i, 1-4; Koch and Nelson, 2002)

Boyd Co.

Lynch Site, 25BD1, 98°28'45"W, 0.75 mi. NW Lynch; age—AD 1473-1511 (Freed, 1954: 7-9; Witty, 1962: 10-13, 136-140)

25BD124, Sec. 13, T34N, R10W, north end Sunshine Bottoms; age—AD 0-1000 (Teter, 1975: 67-68)

Ponca Creek Reservoir, 25BD201, Sec. 4, T34N, R13W, 2.5 mi. NW Butte; age—AD 1000-1500; other designations for the site include Village Site

25BD1 and Anoka Site (White and Cooper, 1951: 28-31)

Ponca Creek, Sec. 5, T24N, R13W, 3.5 mi. NW Butte; age—AD 1000-1400 (White, 1951: 28)

Brown Co.

McIntosh Site, 25BW15, Enders Lake, 42.2639°N, 100.1062°W, ≈12 mi. S, 12.5 mi. W Ainsworth; age—AD 1200-1450 (Koch, 1995: 39-40; 2004: 11-12)

25BW249, 17 mi. N, 6.5 mi. W Johnstown; age—unknown (Pepperl and Falk, 1983: A18-A20)

25BW250, 16.5 mi. N, 6 mi. W Johnstown; age—unknown (Pepperl and Falk, 1983: A20-A23)

25BW252, 16 mi. N, 5.5 mi. W Johnstown; age—unknown (Pepperl and Falk, 1983: A23-A24)

25BW280, 14.5 mi. N, 2 mi. W Johnstown; age—unknown (Pepperl and Falk, 1983: A64-A72)

Buffalo Co.

25BF2, Sec. 17, T10N, R17W, 0.5 mi. NW Amherst; age—unknown (Henning, 1977: 94; Ludwickson, 1978b: 34)

Post Draw Site, 25BF101, Sec. 35, T11N, R13W; age—AD 1000-1350 (Henning, 1977: 23-24; Ludwickson, 1978b: 34)

25BF105, Sec. 10, T10N, R13W; age—AD 1000-1350 (Henning, 1977: 26-27, Ludwickson, 1978b: 34)

25BF109, Sec. 4, T9N, R16W; age—AD 1000-1350 (Henning, 1977: 29-30; Ludwickson, 1978b: 34)

25BF128, Sec. 11, T9N, R16W; age—AD 850-1450 (Henning, 1977: 41-42; Ludwickson, 1978b: 34)

25BF139, Sec. 10, T10N, R13W; age—AD 850-1450 (Henning, 1977: 48; Ludwickson, 1978b: 34)

25BF140, Sec. 35, T11N, R13W; age—AD 850-1450 (Henning, 1977: 48-49; Ludwickson, 1978b: 34)

25BF142, Sec. 16, T9N, R15W; age—AD 1000-1350 (Henning, 1977: 49-50; Ludwickson, 1978b: 34)

Lady Jayne Site, 25BF145, Sec. 11, T9N, R15W; age—AD 1000-1350 (Bleed, 1973: 2)

25BF148, Sec. 7, T9N, R14W; age—AD 1000-1350 (Henning, 1977: 52-53; Ludwickson, 1978b: 34)

25BF153, Sec. 6, T11N, R18W, 0.5 mi. NW Miller; age—AD 1000-1350 (Henning, 1977: 64-65; Ludwickson, 1978b: 34)

25BF155, Sec. 8, T11N, R18W, 0.125 mi. N Miller; age—AD 1000-1350 (Henning, 1977:65-66; Ludwickson, 1978b: 35)

25BF159, Sec. 31, T11N, R17W, 2.25 mi. NW Amherst; age—unknown (Henning, 1977: 67-68; Ludwickson, 1978b: 35)

25BF161, Sec. 18, T10N, R17W; age—unknown (Henning, 1977: 68-69; Ludwickson, 1978b: 35)

Trampec's Bend Site, 25BF171, Sec. 21, T10N, R17W, 1.5 mi. SE Amherst; age—AD 0-1000 (Bleed, 1973: 102, 121)
25BF173, T11N, R18W, 3.5 mi. E Miller; age—AD 1000-1350 (Henning, 1977: 75-76; Ludwickson, 1978b: 35)
25BF178, Sec. 6, T10N, R17W; age—AD 1000-1350 (Henning, 1977: 78-79; Ludwickson, 1978b: 35)
25BF187, Sec. 36, T11N, R18W, 3.75 mi. NW Amherst; age—AD 1000-1350 (Henning, 1977: 82-83; Ludwickson, 1978b: 35)
25BF189, Sec. 25, T11N, R18W, 5.5 mi. NW Amherst; age—AD 1000-1350 (Henning, 1977: 83-84; Ludwickson, 1978b: 35)
25BF193, Sec. 17, T10N, R17W, 0.75 mi. W Amherst; age—AD 1000-1350 (Henning, 1977: 85-86; Ludwickson, 1978b: 35)
25BF199, Sec. 25, T11N, R18W, 4.5 mi. NW Amherst; age—AD 1000-1350 (Henning, 1977: 89; Ludwickson, 1978b: 35)
Flat Rock Site, 25BF210, Sec. 31, T10N, R16W; age—AD 1250-1350 (Bleed, 1973: 52, 101)

Burt Co.
Logan Creek Site A, 25BT3, 41°48.188'W, 96°28.825'W, Logan Creek, SW Oakland; age—BC 4020±160 (Widga, 2006: 65-69, 319)
Logan Creek Site B, 25BT3, 41°48.188'W, 96°28.825'W, Logan Creek, SW Oakland; age—BC 4340±120 (Kivett, 1959; Graham et al., 1987: 240; Widga, 2006: 65-69, 319)
Logan Creek Site C, 25BT3, 41°48.188'W, 96°28.825'W, Logan Creek, SW Oakland; age—BC 5020±90 (Kivett, 1959; Graham et al., 1987: 240; Widga, 2006: 65-69, 319)
Logan Creek Site D, 25BT3, 41°48.188'W, 96°28.825'W, Logan Creek, SW Oakland; age—BC 5070±110 (Widga, 2006: 65-69, 319)
Logan Creek Site F, 25BT3, 41°48.188'W, 96°28.825'W, Logan Creek, SW Oakland; age—BC 5350±270 (Widga, 2006: 65-69, 319)
Logan Creek Site G, 25BT3, 41°48.188'W, 96°28.825'W, Logan Creek, SW Oakland; age—unknown (Widga, 2006: 65-69, 319)
25BT11, 3.5 mi. E Bancroft; age—unknown (Carlson, 1973: 9-10)

Butler Co.
Linwood Site, 25BU1, Skull Creek, 1 mi. SW Linwood; age—AD 1725-1775 (Carlson, 1973: 57-80)
Barcal [Skull Creek] Site, 25BU4, near Abie; age—AD 1700-1750 (O'Shea and Ludwickson, 1992: 338)
White Site, 25BU20; age—AD 0-1000 (Ore, 1981: 1)
Palmer Johnson Site, 25BU37, east of Ulysses; age—AD 1275-1300 (Souders, 1994: 45, 53; Bozell and Koch, n.d.)

Cass Co.
Ashland Site, 25CC1, ≈ 3 mi. E Ashland [in Saunders Co.]; age—Occupations A and C, around AD 1650-1750 (Hill and Cooper, 1937: 249-252)
Ashland Site, 25CC1, ≈ 3 mi. E Ashland [in Saunders Co.]; age—Occupation B, AD 1000-1400 (Hill and Cooper, 1937: 249-252)
Theodore Davis Site, 25CC17, Sec. 6, T10N, R12E, 1.5 mi. E Weeping Water; age—AD 1200-1345 (Gradwohl, 1969: 63, 115; Blakeslee and Caldwell, 1979: 1-22)
Walker Gilmore, 25CC28, 6 mi. SE Murray, Sec. 28, T11N, R14E; age—AD 800-1285 (Sterns, 1915: 122; Strong, 1935: 176; Haas, 1983; Graham et al., 1987: 239; NRHP)
Rock Bluffs Site, 25CC31, 1 mi. S Rock Bluff; age—AD 1000-1400 (Strong, 1935: 124-125; Bell and Gilmore, 1936: 311)
Little Pawnee Creek Site, 25CC228, 0.8 mi. N, 3 mi. W South Bend; age—AD 1287-1409 (Bozell and Ludwickson, 1994:100-103)
Behrnes Site, Sec. 14, T10N, R13E; age—AD 1000-1400 (Bell and Gilmore, 1936: 312)
Cornish Site, near Louisville; age—AD 1000-1400 (Hill and Cooper, 1937: 287-290)
Gates Site, 2 mi. W Fort Crook, Bellevue; age—AD 1000-1400 (Bell and Gilmore, 1936: 311)
North Plattsmouth Site, N of Plattsmouth and S of Platte River; age—AD 1000-1350 (Hill and Cooper, 1937: 311-314)
Sheldon Site, Sec. 21, T10N, R13E; age—AD 1000-1400 (Bell and Gilmore, 1936: 312)
Woolsey Site, Sec. 5, T10N, R14E; age—AD 1000-1400 (Bell and Gilmore, 1936: 312)

Cedar Co.

Schulte Site, 25CD1, St. Helena; age—AD 1350-1450 (Cooper, 1936: 15-19)
Radke Site, 25CD2, 3 mi. S Wynot; age—AD 1350-1450 (Cooper, 1936: 15-19)
Wiseman Site, 25CD3, 4.5 mi. NE Wynot; age—AD 1350-1450 (Cooper, 1936: 15-16, 19-21)
Ferber Site, 25CD10, near Bow Valley Mill; age—AD 800-1300 (Ludwickson et al., 1981: 133-140)
25CD36, East Bow Creek, 4.5 mi. S, 1.25 mi. E Wynot; age—BC 3000-AD 500 (Watson et al., 2004)

Chase Co.

Lovitt Site, 25CH1, 12 mi. W Wauneta; age—circa AD 1700 (Hill and Metcalf, 1942: 158-164; Gunnerson, 1960: 212-216)
Ough Site, 1 mi. E Wauneta; age—AD 0-1000 (Hill and Kivett, 1940: 232)

Cherry Co.

Barrow Pit Site, 25CE231, Sec. 22, T34N, R27W, Fort Niobrara Wildlife Refuge; age—unknown (Osborn, 1979)
25CE255, 3 mi. S, 2 mi. W Sparks; age—unknown (Pepperl and Falk, 1983: A120)
25CE309, 3 mi. S, 0.4 mi. W Sparks; age—AD 0-1000 (Pepperl and Falk, 1983: A221-A225)
25CE312, 3.5 mi. S, 1 mi. W Sparks; age—AD 0-1650 (Pepperl and Falk, 1983: A229-A246)

Cheyenne Co.

Thurston Site, 25CN11, Lodgepole Creek, 6 mi. E Potter; age—AD 1-1000 (Jensen, 1973: 165-167)
Rock Shelter Site, 25CN23, 4 mi. E Potter; age—unknown (Jensen, 1973: 201)
Cave Site, 25CN24, 4 mi. E Potter; age—AD 1000-1350 (Jensen, 1973: 201-202)
25CN26, Point of Rocks, 4 mi. E Potter; age—unknown (Jensen, 1973: 171)
Rock Shelter Site, 25CN27, 4 mi. E Potter; age—AD 1000-1350 (Jensen, 1973: 202-203)
Fred Carlson Site, 25CN29, Lodgepole Creek, 3 mi. E Potter; age—AD 1675 (Jensen, 1973: 171-176)
25CN55, southeastern Sidney; age—pre-1700 (Carlson, 1999: 105-108)
25CN56, Lodgepole Creek, 1 mi. S Potter; age—BC 7500-7000 (Steinacher and Carlson, 1984: 25-28)

Colfax Co.

Schuyler Site [Gray-Wolfe Site], 25CX1, along Shell Creek, ≈ 3 mi. NW Schuyler; age—AD 1500-1650 (Dunlevy, 1936: 154-157; Bozell et al., 1982: 1-2; Graham et al., 1987: 240)

Custer Co.

Dickman Ossuary, 25CU7, Sec. 24, T15N, R22W; age—AD 1000-1350 (Philips, 1960a: 1-2)
"Doc" McKenzie Site, 25CU92, Spring Creek, 4.6 mi. S, 0.25 mi. E Comstock; age—BC 6200-5500 (Holen and Watson, 1999: 82-101)
25CU101, Middle Loup River valley, 4.75 mi. S, 0.75 mi. E Comstock; age—AD 1550-1750 (Holen and Watson, 1999: 52-61)

Dakota Co.

Bobier Site, 25DK1, 4.5 mi. SE Homer; age—AD 1450-1500 (Frantz, 1963: 2, 149-151)
25DK3, Sec. 23, T27N, R7E, 6 mi. W Homer; age—AD 0-1000 (Price, 1956: 90-92, 103-107)
Big Village Site, 25DK5, 1 mi. NE Homer; age—AD 1775-1845 (Steinacher and Carlson, 1984: 29-36; O'Shea and Ludwickson, 1992: 1-6)
Nelson Site, 25DK7, 1 mi. N Homer; age—AD 1450-1500 (Frantz, 1963: 2, 149-151)
25DK10, 1 mi. N Homer; age—AD 1775-1845 (O'Shea and Ludwickson, 1992: 1-6)
Hancock Site, 25DK14, 1 mi. SE Homer; age—AD 1450-1500 (Frantz, 1963: 3, 149-151)

Dawes Co.

Slaughterhouse Creek Site, 25DW17, western edge of Crawford; age—AD 1550-1700 (Bozell and Ludwickson, 1988: 1-3, 21-22)
25DW19, 0.75 mi. W Crawford; age—BC 1000-AD 100 and post-1700 (Bozell and Ludwickson, 1988: 1-3, 69-73)
25DW21, 0.6 mi. S, 2 mi. W Crawford; age—AD 400-1000 (Bozell and Ludwickson, 1988: 1-3, 73-76)
25DW59, 0.70 mi. W Crawford; age—BC 1000-AD 900 (Bozell and Ludwickson, 1988: 1-3, 76-88)
Big Cottonwood #3, 25DW94, Sec. 9, T33N, R52W; age—unknown (Meston, 197625)
25DW102, Sec. 31, T33N, R52W; age—BC 7000-6000 (Meston, 1976: 31-32)

FN-2, 2.5 mi. S, 3 mi. E Crawford; age—unknown (Bozell and Pepperl, 1987: 23, 69)

Dawson Co.
25DS7, Sec. 5, T11N, R19W; age—AD 1000-1350 (Henning, 1977: 113-114; Ludwickson, 1978b: 35)
25DS19, Sec. 3, T8N, R20W, right of way for I-80 near Platte River; age—AD 1550-1700 (Frantz, 1965: 20)
25DS21, Platte River valley, 4.75 mi. S, 4 mi. E Lexington; age—AD 1600 (Garrett, 1965: 75-76; Roper, 1994: 55-95)
25DS102, Sec. 36, T10N, R19W; age—AD 1000-1350 (Henning, 1977: 100-101)
25DS106, Sec. 26, T10N, R19W; age—AD 1000-1350 (Henning, 1977: 103; Ludwickson, 1978b: 36)
25DS107, Sec. 6, T11N, R19W, 1.25 mi. W Sumner; age—AD 1000-1350 (Henning, 1977: 104; Ludwickson, 1978b: 36)
25DS108, Sec. 1, T11N, R19W; age—AD 1000-1350 (Henning, 1977: 104-105; Ludwickson, 1978b: 36)
25DS109, Sec. 4, T8N, R20W, 6 mi. SW Overton; age—AD 1000-1350 (Henning, 1977: 105; Ludwickson, 1978b: 36)
25DS110, Sec. 2, T11N, R19W; age—AD 1000-1350 (Henning, 1977: 106; Ludwickson, 1978b: 32)
25DS111, Sec. 33, T12N, R20W; age—AD 1000-1350 (Henning, 1977: 106-107; Ludwickson, 1978b: 36)
25DS112, Sec. 36, T12N, R20W; age—AD 1000-1350 (Henning, 1977: 107; Ludwickson, 1978b: 36)
25DS117, Sec. 16, T10N, R19W; age—AD 1000-1350 (Henning, 1977: 109-110; Ludwickson, 1978b: 36)
25DS118, Sec. 16, T10N, R19W; age—AD 1000-1350 (Henning, 1977: 110-111; Ludwickson, 1978b: 36)
25DS120, Sec. 1, T11N, R20W; age—AD 1000-1350 (Henning, 1977: 112; Ludwickson, 1978b: 36)

Deuel Co.
Neumann Site, 25DU3, Lodgepole Creek, 3.5 mi. SE Chappell; age—AD 1-1000 (Carlson, 1973: 101-105)

Dixon Co.
Peterson Site, 25DX1, near Ponca; age—AD 1350-1450 (Hill, 1988: 99-132)
25DX2, Sec. 5, T29N, R6E, 3 mi. S, 2 mi. W Ponca; age—AD 0-1000 (Price, 1956: 108-111, 123-125)
Annie's Site, 25DX30, in vicinity of 2 mi. N, 3 mi. E Newcastle; age—AD 1410-1480 (Blakeslee, 1988: 1-12)

Douglas Co.
Champe Site, 25DO1, margin of uplands overlooking bottom lands through which flow the Elkhorn and Platte rivers; age—approximately AD 1300-1400 (Hill and Cooper, 1936b: 253-254, 266-268; NRHP)
Parker Site, 25DO2, on the bluffs north of Ponca Creek and west of the Missouri River, in extreme northeastern Douglas Co.; age—AD 1000-1350 (Ewing, 2000: 1-2)
25DO3, Ponca Creek in northeastern corner of county; age—AD 1000-1400 (Cooper, 1939: 119)
Saunders Site, 25DO5, eastern side of Elkhorn River in southwestern corner of county; age—AD 1000-1400 (Strong, 1935: 168-170)
Cabanne's Post, 25DO8, near Dodge Park northeast of Omaha; age—AD 1822-1839 (Bozell et al., 1990: 1)
25DO11, 2.5 mi. W Florence; age—AD 110-1400 (Steinacher and Carlson, 1984: 37)
Long's Hill, 25DO26, on Ponca Creek in extreme northeastern corner of county; age—AD 1000-1400 (Gilder, 1908: 60-61, 70)
Ponca Creek District, 2.25 mi. N former town of Florence; age—unknown (Gilder, 1907: 702-703)
unnamed site, approximately 15th and I Streets, Omaha; age—unknown (Gilder, 1909: 78)

Franklin Co.
Lost Creek Site [Dooley Site], 25FR6, Lost Creek, 2.25 mi. S, 1.5 mi. E Bloomington; age—AD 1000-1350 (Strong, 1935: 69-73)
Holdrege 3, head of Lost Creek, 6 mi. S Bloomington; age—unknown (Wedel, 1934: 157)
Holdrege 4, Rebecca Creek, 6 mi. SW Bloomington; age—unknown (Wedel, 1934: 161)
Red Cloud 2, 4 mi. SE Franklin; age—unknown (Wedel, 1934: 149)
Red Cloud 3, 3.5 mi. S, 1.5 mi. E Franklin; age—unknown (Wedel, 1934: 152)

Frontier Co.
Owens Site, 25FT3, on Medicine Creek; age—AD 1000 (Mick, 1983: 70; Graham et al., 1987: 240)
Dick Site [= Medicine Creek 5], 25FT9, between the confluence of Brush Creek and Medicine Creek, 2 mi. W Curtis; age—circa AD 1700 (Gunnerson 1960: 222)

25FT13, vicinity of Medicine Creek Dam; age—AD 1010 (Kivett and Metcalf, 1997: 45-56; Mick, 1983: 71; Graham et al., 1987: 240)

25FT14, vicinity of Medicine Creek Dam; age—AD 1250-1400 (Mick, 1983: 71; Kivett and Metcalf, 1997: 56-58)

25FT16, vicinity of Medicine Creek Dam; age—AD 1020-1235 (Mick, 1983: 71; Graham et al., 1987: 240; Kivett and Metcalf, 1997: 58-59)

25FT17, vicinity of Medicine Creek Dam; age—AD 1080-1240 (Mick, 1983: 72; Graham et al., 1987: 241-242; Kivett and Metcalf, 1997: 11-14, 32-45)

25FT18, near point where Lime Creek originally entered Medicine Creek; age—AD 595±225 (Kivett and Metcalf, 1997: 14-19)

25FT19, near point where Lime Creek originally entered Medicine Creek; age—AD 1000-1350 (Kivett and Metcalf, 1997: 60-63)

25FT20, near point where Lime Creek originally entered Medicine Creek; age—AD 1000-1350 (Kivett and Metcalf, 1997: 63-64)

25FT22, ≈ 4 km upstream on Medicine Creek from Medicine Creek Dam, or Sec. 11, T5N, R26W; age—AD 1100-1400 (Turnmire, 1996: 189-191; Kivett and Metcalf, 1997: 64-68)

25FT28, near point where Lime Creek originally entered Medicine Creek; age—AD 1000-1350 (Kivett and Metcalf, 1997: 68-70)

25FT30, near point where Lime Creek originally entered Medicine Creek; age—AD 1100-1300 (Mick, 1983: 72-73; Graham et al., 1987: 242; Kivett and Metcalf, 1997: 70-72)

Spring Creek Site, 25FT31, Red Willow Creek, adjacent to Red Willow Dam, Sec. 36, T5N, R30W; age—BC 3850-3500 (Kivett, 1961: 7; Grange, 1980: 12-47; Widga, 2004: 25-30)

25FT32, Red Willow Creek, 0.75 mi. upstream from Red Willow Dam; age—AD 1362 (Grange, 1980: 49-69; Mick, 1983: 68)

Mowry Bluff Site, 25FT35, Sec. 25, T5N, R26W, 6 mi. NNW Cambridge [in Furnas Co.] ; age—AD 1020-1180 (Wood, 1969a: 3, 1969b: 14; Mick, 1983: 73; Graham et al., 1987: 242)

25FT36, ≈ 4 km upstream on Medicine Creek from Medicine Creek Dam; age—AD 1175±25 (Mick, 1983: 73-74; Kivett and Metcalf, 1997: 75-76)

25FT39, ≈ 5 km upstream on Medicine Creek from Medicine Creek Dam; age—AD 1200-1280 (Mick, 1983: 74; Graham et al., 1987: 242; Kivett and Metcalf, 1997: 76-82)

Lime Creek Site, 25FT41, north side of Lime Creek valley, 1 mi. from original junction of Lime and Medicine creeks; age—BC 7500-6000 (Davis, 1962: 23; Graham et al., 1987: 243; Wedel, 1986: 66-71; Bamforth, 2002: 58)

Red Smoke Site, 25FT42. north side of Lime Creek valley, 1.3 mi. from original junction of Lime and Medicine creeks; age—BC 8000-6000 (Davis, 1962: 23; Wedel, 1986; 66-71; Bamforth, 2002: 58)

Allen Site, 25FT50, ≈ 3 mi. upstream from Medicine Creek Dam; age—BC 11,000-7500 (Holder and Wike, 1949: 260; Wedel, 1986: 66-71; Bamforth, 2002: 58; 2007: 1-8)

25FT54, Red Willow Creek, 5 mi. upstream from Red Willow Dam; age—AD 1310 (Grange, 1980: 87-93; Mick, 1983: 68-69)

25FT58, Red Willow Creek, ≈4.5 mi. upstream from Red Willow Dam; age— AD 1000-1350 (Grange, 1980: 135-138)

25FT70, vicinity of Medicine Creek Dam; age—AD 690-1450 (Mick, 1983: 74-75; Graham et al., 1987: 242-243; Kivett and Metcalf, 1997: 19- 22, 82-93)

25FT77, Red Willow Creek, ≈1 mi. upstream from Red Willow Dam; age—AD 600±250 (Grange, 1980: 129-132)

25FT80, Red Willow Creek, 4.5 mi. upstream from Red Willow Dam; age—AD 1476 (Grange, 1980: 71-86; Mick, 1983: 69)

25FT177, vicinity of Medicine Creek Reservoir; age—Late Pleistocene, BC 15,000 (Holen and May, 1989: 1-3)

Medicine Creek 4, 2 mi. S Stockville; age—unknown (Wedel, 1934: 166)

Garden Co.

Ash Hollow Cave Site, 25GD2, 3 mi. SE Lewellen; age—AD 0-1700 (Champe, 1946: 54-57)

O. V. Clary site, 25GD50, Ash Hollow Draw, 4.25 mi. S, 1 mi. W Lewellen; age—BC 7000 (Hill et. al., 2008: 115-130)

25GD54, Blue Creek, 2 mi. N, 9 mi. E Oshkosh; age—prehistoric (Bozell et al., 2008: 59)

25GD84, Blue Creek, 2.5 mi. N, 9.75 mi. E Oshkosh; age—prehistoric (Bozell et al., 2008: 82)

Clary Ranch Site, 25GD106, Ash Hollow Draw, 4.25 mi. S Lewellen; age—BC 7000 (Hill, 2005: 249-254; Hill et. al., 2008: 115-130)

Barn Bluff, 6.5 mi. SE Oshkosh; age unknown (Renaud, 1933: 23-24, 48)
unnamed site, North Platte River, 4.5 mi. W Oshkosh; age—unknown (Renaud, 1933: 19, 48)

Garfield Co.
25GF13, 1 mi. S, 2.1 mi. W Burwell; age—AD 1-400 (Bozell and Pepperl, 1982: 2-13)
25GF14, 1 mi. S, 1.5 mi. W Burwell; age—AD 1250-1450 (Bozell and Pepperl, 1982: 13-19)

Gosper Co.
Wallace Site, 25GO2, Plum Creek, 5.5 mi. N, 9.5 mi. E Elwood; age—AD 425-650 (Winfrey, 1991: 58-64; Souders, 1994: 44-45, 53)
Mousel Site (Turkey Creek I), 25GO5, Sec. 36, T5N, R22W, 16 mi. S, 5 mi. W Elwood; age—AD 1000-1350 (Philips, 1957a: 1)
Scheivalbein Site, 25GO21, Sec. 26, T5N, R23W, 3 mi. W, 16 mi. S Elwood; age—AD 1000-1350 (Philips, 1954: 1)
Golter Site, 25GO30, Sec. 35, T5N, R23W, 4 mi. N Arapahoe [Furnas Co.]; age—AD 1000-1350 (Philips, 1957b: 1)
Stillman Site, 25GO35, Sec. 24, T5N, R24W, about 15 mi. S, 8 mi. W Elwood; age—AD1000-1350 (Philips, 1960b: 1)
Helms Site, 25GO39, Sec. 14, T5N, R24W, 14 mi. S, 9 mi. W Elwood; age—AD 1000-1350 (Philips, 1956a: 1)
Graf Site, 25GO44, Sec. 11, T5N, R24W, 13 mi. S, 9 mi. W Elwood; age—AD 1000-1350 (Phillips, 1957c: 1)
Greisfeller Site, 25GO50/51, 13 mi. S, 10 mi. W Elwood; age—AD 1000-1350 (Philips, 1960c: 1)
Reuben Lucas Site, 25GO52, Sec. 3, T5N, R24W, 12 mi. S, 10 mi. W Elwood; age AD 1000-1350 (Philips, 1958: 1)
Gross Site, 25GO53, Sec. 3, T5N, R24W, 12 mi. S, 10 mi. W Elwood; age—AD 1000-1350 (Philips, 1956b: 1)
Lowell Dawson Site, 25GO87, Sec. 1, T5N, R22W; age—AD 850-1450 (Philips, 1959: 1)
Dawson Site, 25GO88, Sec. 12, T5N, R22W; age—AD 850-1450 (Philips, 1960d: 1)

Greeley Co.
Schudel Site, 25GY12, North Loup River valley, vicinity Scotia; age—BC 6000-3000 (Roper, 1998: 1-34)

Hall Co.
25HL3-25HL4, Sec. 29, T9N, R12W, right of way for I-80 near Platte River; age—AD 1550-1700 (Frantz, 1965: 40)
Hulme Site, 25HL28, 16 mi. W Grand Island; age—AD 1170-1220 (Bozell, 1991: 229-231; Souders, 1994: 45, 53)

Harlan Co.
Graham Ossuary, 25HN5, 4.25 S, 4.5 E Alma; age—AD 1050-1350 (Adair and Brown, 1987: 78-79)
Graham Site, Prairie Dog Creek, 2 mi. S, 0.5 mi. W Republican City; age—AD 1000-1350 (Strong, 1935: 103-105)
25HN12, 6 mi. S Alma; age—AD 400-1350 (Adair and Brown, 1987: 95-100)
Sindt Point, 25HN16, 3 mi. S, 2.75 mi. E Alma; age—unknown (Adair and Brown, 1987: 111-115)
Stevenson Village, 25HN31, 1.5 mi. S, 0.25 mi. W Alma; age—mixed sample (Adair and Brown, 1987: 118-123)
25HN33, 1.75 S mi. Alma; age—AD 1050-1350 (Adair and Brown, 1987: 135-136)
25HN36, 2 mi. S Alma; age—AD 1050-1350 (Adair and Brown, 1987: 144-154)
White Cat Village, 25HN37, Sec. 24, T1N, R18W, Prairie Dog Creek, 6 mi. SE Alma; age—AD 1675-1720 (Champe, 1949: 285; Gunnerson, 1960:145-146; Adair and Brown, 1987: 161-168)
Green Plum Site, 25HN39, 5.5 mi. S, 0.8 mi. E Alma; age—AD 1500-1600 (Rusco, 1960: 3; Adair and Brown, 1987: 181-182)
25HN40, 0.8 mi. S, 2 mi. E Alma; age—AD 400-900 (Adair and Brown, 1987: 188-194)
Indian Hill, 25HN42, 3.75 mi. S, 4 mi. E Alma; age—AD 1050-1350 (Adair and Brown, 1987: 203-207)
Blue Stone Site, 25HN45, Prairie Dog Creek, 7 mi. SE Alma, Sec. 23, T1N, R18W; age—AD 1500-1600 (Rusco, 1960: 11)
Rural School District No. 9, 25HN50, 0.8 mi. S, 1.75 mi. E Alma; age—AD 1500-1750 (Adair and Brown, 1985: 211-213)
25HN54, 3 mi. S, 2.5 mi. E Alma; age—AD 1050-1350 (Adair and Brown, 1987: 234-237)
25HN57, 4.5 mi. S, 3 mi. E Alma; age—mixed site (Adair and Brown, 1987: 240-243)
25HN62, 4.5 mi. S, 2 mi. E Alma; age—unknown (Adair and Brown, 1987: 258-262)

25HN64, 0.75 S, 2.5 mi. W Alma; age—AD 1000-1750 (Moore, 1988: 38-40)
25HN65, 1.25 mi. W Alma; age—AD 1000-1750 (Moore, 1988: 40-48)
25HN125, 0.75 mi. S, 2.1 mi. E Alma; age—AD 400-900 (Adair and Brown, 1987: 273-276)
Tipover Cove Site, 25HN178, 1.25 mi. S, 3 mi. E Alma; age—BC 700-AD 550 (Holen et al. 1996: 46-49)
25HN180, 2 mi. S, 3.5 mi. E Alma; age—BC 8500-7700 (Holen et al., 1996: 15)
25HN182, 4.2 mi. S, 5.9 mi. E Alma; age—BC 8500-7700 (Holen et al., 1996: 15)

Hitchcock Co.
Carmody Site, 25HK7, 2 mi. W Trenton; age—AD 500-1000 (Kivett, 1951: 42-45; Kivett, 1952: 24-30)
Massacre Canyon Site, 25HK13, 40º11'33"N, 100º57'46"W, ≈ 3 mi. E Trenton; age—AD 500-1000 (Kivett, 1951: 8-12; Kivett, 1952: 5-10)

Holt Co.
25HT1, Sec. 4, T31N, R12W, 17 mi. N, 3 mi. W O'Neill; age—AD 0-1000 (Price, 1956: 1-9)
25HT2, Sec. 14, T32N, R10W, Redbird Creek, 22 mi. NE O'Neill; age—AD 1600-1700 (Wood, 1956: 54, 67; 1965: 85-86)
Redbird Site, 25HT3, Sec. 12, T32N, R10W, 0.2 mi. N Redbird; age—AD 1600-1700 (Wood, 1956: 5, 52; 1965: 86-90)

Hooker Co.
Lowe Site, 25HO7, 1.5 mi. N, 1 mi. E Mullen; age—circa AD 1700 (Gunnerson, 1960: 145, 183-184)
Humphrey Site [= Matthews Site and Mullen Reservoir], 25HO21, Sec. 18, T24N, R31W, Middle Loup River, about 5 mi. E Mullen; age—circa AD 1700 (Gunnerson, 1960: 187-205; Kivett, 1947: 28)
Kelso Site, 25HO23, Middle Loup River, about 5 mi. E Mullen; age—AD 0-1000 (Kivett, 1952: 33-35)

Howard Co.
Palmer Site, 25HW1, Sec. 12, T15N, R9W, 4 mi. N, 1 mi. W Palmer; age—post-AD 1700 (Wedel, 1936: 25-26)
Stabaco Site, 25HW16, Loup River, near Cushing; age—AD 1725-1760 (Holen, 1995: 1-13; Holen and Peterson, 1995: 219-220)
Schmidt Site, 25HW301, Sec. 13, T15N, R11W, along North Loup River, near Elba; age—AD 1100-1550 (Satorius-Fox, 1982: 1-4; Mick, 1983: 76; Graham et al., 1987: 240)
unnamed site, Munson Creek, 3 mi. W Elba; age—unknown (Hill, 1932: 172-175)

Keith Co.
25KH17, Sec. 11, T13N, R39W; age—circa AD 1700 (Gant, 1966: 3)
25KH73, Sec. 13, T13N, R37W; age—BC 4000 (Holen and Watson, 2000: 67)
25KH83, Brule Canyon, near Brule; age—pre-AD 1700 (Holen and Watson, 2000: 73)

Keya Paha Co.
25KP243, East Middle Creek, 2.25 mi. S, 1.25 mi. E Norden; age—AD 850-1450 (Pepperl and Falk, 1983: A295-A310)
25KP261, 0.5 mi. S, 2.5 mi. W Norden; age—unknown (Pepperl and Falk, 1983: A333-A335)
25KP311, 5.25 mi. S, 1 mi. E Norden; age—unknown (Pepperl and Falk, 1983: A382-A385)

Knox Co.
Ponca Fort, 25KX1, Sec. 29, T33N, R7W, 1 mi. E Verdel; age—AD 1715 (Wood, 1993: 3-4, 79)
Minarik Site, 25KX2, Sec. 29, T33N, R7W, Ponca Creek, 7 mi. NW Verdel; age--post-AD 1700 (Wood, 1956: 84, 141)
Wirth Site, 25KX4, Sec. 2, T32N, R8W, Niobrara River, 14 mi. upstream from confluence of Niobrara and Missouri rivers; age—AD 1600-1700 (Wood, 1956: 68, 83; 1965: 90-91)
Minarik Site, 25KX9, Sec. 29, T33N, R7W, Ponca Creek, 7 mi. NW Verdel; age—AD 1650-1750 (Wood, 1956: 84, 129; 1965: 91-94)
25KX15, Sec. 15, T33N, R3W, about 4 mi. N, 19.5 mi. Niobrara; age—AD 750-1000 (Blakeslee and O'Shea, 1983: 177-179)
25KX30, Sec. 13, T33N, R5W, Santee; age—AD 1300-1450 (Carlson, 1976: 4)

25KX52, Sec. 17, T33N, R2W, east bank Weigand Creek, about 4 mi. N, 24 mi. E Niobrara; age—AD 800-1260 (Blakeslee and O'Shea, 1983: 179-181)
25KX53, Sec. 17, T33N, R2W, near entrance Weigand Creek Recreation Area, about 4 mi. N, 24 mi. E Niobrara; age—unknown (Blakeslee and O'Shea, 1983: 181-183)
25KX56, Sec. 20, T33N, R3W, mouth of Devils Nest Creek, about 3.75 mi. N, 15 mi. E Niobrara; age—BC 50-AD 500 (Blakeslee and O'Shea, 1983: 187-191)
25KX57, Sec. 17, T33N, R2W, mouth Weigand Creek, about 4.5 mi. N, 24 mi. E Niobrara; age—AD 800-1500 (Blakeslee and O'Shea, 1983: 191-199)
25KX59, Sec. 14, 13, 18, T33N, R2-3W, near Bloomfield Recreation Area, about 4.5 mi. N, 21 mi. E Niobrara; age—unknown (Blakeslee and O'Shea, 1983: 203-204)
25KX64, Sec. 22, T33N, R5W, about 3 mi. N, 7.5 mi. E Niobrara; age—unknown (Blakeslee and O'Shea, 1983: 209)
25KX67, Sec. 27, T33N, R5W, about 2.5 mi. N, 6.5 mi. Niobrara; age—BC 50-AD 500 (Blakeslee and O'Shea, 1983: 213-215)
25KX68, Sec. 33, T33N, R5W, about 2 mi. N, 6.25 mi. E Niobrara; age—unknown (Blakeslee and O'Shea, 1983: 215-217)
25KX69, Sec. 33, T33N, R5W, about 2 mi. N, 6.25 mi. E Niobrara; age—unknown (Blakeslee and O'Shea, 1983: 217-218)
25KX71, Sec. 13, T32N, R6W, about 3 mi. E Niobrara; age—AD 0-1000 (Blakeslee and O'Shea, 1983: 220-223)
25KX200, Sec. 18, T33N, R2W, about 4.5 mi. N, 23 mi. E Niobrara; age—AD 0-1450 (Frantz, 1965: 50; Blakeslee and O'Shea, 1983: 227-237)
25KX203, Sec. 15, T33N, R2W, Deepwater Recreation Area, about 4.5 mi. N, 25 mi. E Niobrara; age—AD 0-1000 (Blakeslee and O'Shea, 1983: 239-242)
Tramp Deep Site, 25KX204, Sec. 14, T33N, R2W, about 5 mi. E Niobrara; age—BC 6000-AD 1000 (Howard and Gant, 1966: 16-17; Blakeslee and O'Shea, 1983: 242-245)
Niobrara Railroad Bridge Site, 25KX207, Sec. 7, T33N, R6W, confluence of Niobrara and Missouri rivers; age—circa 1850 (Howard and Gant, 1966: 24-27)

Lancaster Co.
Schrader Site, 25LC01, Salt Creek, near Roca, about 9 mi. S Lincoln; age—AD 1000-1500 (Hill and Cooper, 1936a: 223-224, 246-248; Blakeslee and Caldwell, 1979: 1-22)

Lincoln Co.
Keith Ranch Site, 25LN34, abandon Keith Ranch, south of South Platte River; age—AD 1675-1725 (Carlson, 1973: 32)

Loup Co.
Blowout Site, 25LP7, 0.75 mi. upstream on Dry Creek from its confluence with Calamus River; age—AD 1650-1750 (Falk and Pepperl, 1981: A1 41-52)
Royal Goodenow Site, 25LP8, confluence of Dry Creek and Calamus River, 8 mi. NW Burwell [Garfield Co.]; age—AD 1500-1750 (Falk and Pepperl, 1981: A1 52-93; Roper, 1989: 4)
Jane Toben Site, 25LP11, 0.5 mi. N confluence of Dry Creek and Calamus River; age—BC 1000-AD 1000 (Falk and Pepperl, 1981: A1 93-106)
Gaylord Wallace Site, 25LP14, 0.75 mi. upstream on Gracie Creek from its confluence with Calamus River; age—AD 900-1700 (Falk and Pepperl, 1981: A1 113-124)
Gracie Creek Site, 25LP16, 0.5 mi. upstream on Gracie Creek from its confluence with the Calamus River; age—AD 1850-1910 (Falk and Pepperl, 1981: A1 132-153)
Russel L. Bald Site, 25LP17, 1.5 mi. upstream on Dry Creek from its confluence with the Calamus River; age—AD 1650-1750 (Falk and Pepperl, 1981: A1 153-161)
Eldon Larson Site, 25LP19, ≈ 6.5 mi. upstream from Calamus Dam; age—AD 1650-1750 (Falk and Pepperl, 1981: A1 166-174)

Madison Co.
25MD3, near Union Creek, 1.75 mi., E Madison; age—AD 1000-1150 (Steinacher and Carlson, 1984: 46-59)

McPherson Co.

Dry Lake Burial Site, 25MP2, Sec. 31, T20N, R34W, Dry Lake, 13 mi. NW Tryon; age—BC 100-AD 500 (Oothoudt, 1976: 78-81, 166-167)

Merrick Co.

Tahaksu Site, 25MK15, 4.8 mi. N, 1.2 mi. W Palmer; age—AD 1100-1400 (Peterson and Holen, 1996)

Morrill Co.

25MO14, Lower Dugout Creek Drainage, Sec. 9, T19N, R47W; age—prehistoric (Bozell, 2008: 157-158)
25MO37, Lower Dugout Creek Drainage, Sec. 3, T19N, R47W; age—AD 1550-1750 (Bozell et al., 2008: 160)
25MO169, Greenwood Creek, ≈ 5 mi. SE Bridgeport; age—prehistoric (Bozell et al., 2008: 139-140)
Greenwood Site, old Greenwood Stage Station, Keenan Ranch, ≈ 9 mi. S, 8 mi. E Redington; age—unknown (Renaud, 1933: 12-15)
Site 2, 1 mi. S of head Cedar Creek to the south of Broadwater; age—unknown (Bell and Cape, 1936: 375-376)
Site 21, 2.5 mi. SW Mud Springs; age—unknown (Bell and Cape, 1936: 365-366)
Site 45, ≈ 9 mi. S, 8 mi. E Redington; age—unknown (Bell and Cape, 1936: 367-372)

Nance Co.

Burkett Site, 25NC1, Sec. 20, T17N, R4W, 4 mi. SW Genoa; age—AD 1550-1750 (Blackman, 1907: 339; 1924: 1; Dunlevy, 1936: 151-152; Wedel, 1936: 40)
Horse Creek Site, 25NC2, near Fullerton; age—AD 1809-1839 (Bozell, 1988: 98; Hill, 1927)
Wright Site, 25NC3, near Genoa, AD1650-1750 (Bozell, 1988: 98; Wedel, 1938)
25NC8, vicinity of Fullerton; age—AD 1250-1450 (Ludwickson, 1978: 94-108; Mick, 1983: 76; Graham et al., 1987:240)
25NC10, vicinity of Fullerton; age—AD 1250-1450 (Ludwickson, 1978: 94-108; Mick, 1983: 76; Graham et al., 1987:240)
25NC12, vicinity of Fullerton; age—AD 1250-1450 (Ludwickson, 1978: 94-108; Mick, 1983: 76-77)
25NC13, vicinity of Fullerton; age—AD 1250-1450 (Ludwickson, 1978: 94-108; Mick, 1983: 76-77)
Palmer Locality [in part], 25NC29, 5.5 mi. S, 19 mi. W Fullerton; age—AD 1100-1400 (Holen et al., 2001: 1-4; Peterson and Holen, 2001: 178)
unnamed site, 3 mi. E Genoa; age—AD 1750-1850 (Wedel, 1936: 54, 83-84)

Nemaha Co.

Leahy Site, 25NH1, ≈ 3 mi. W Peru; age—AD 0-1000 (Hill and Kivett, 1940: 199)
Majors Site, 25NH2, ≈ 3 mi. NW Peru; age—AD 950-1300 (Hill and Cooper, 1937: 318-324; Blakeslee and Caldwell, 1979: 1-22)
Osborne Site, 25NH5, Missouri River, N of Brownsville; age—AD 1000-1400 (Hill and Cooper, 1937: 335-338)
McKissick Site, 3 mi. NW Peru; age—AD 0-1000 (Hill and Cooper, 1937: 349; Hill and Kivett, 1940: 199)

Nuckolls Co.

25NO2, Sec. 3, T4N, R7W, terrace about 1/4 mi. Blue River; age—BC 3000-AD 500 (Frantz, 1965: 60)

Otoe Co.

25OT24, Sec. 26, T8N, R14E; age—AD 1000-1450 (Kaldahl, 1993: 106-112) [there is some confusion in this paper about the placement of site; it is not in Nemaha Co. as indicated at some points, but at this place in Otoe Co.]

Pawnee Co.

Table Rock Site, 25PW1, Sec. 33, T3N, R12E; age—AD 1000-1400 (Bell and Gilmore, 1936: 312)

Platte Co.

Bakenhus Site, 25PT2, Loseke Creek about 50 m below forks of Elk and Deer creeks, Sec. 15, T19N, R1E; age—AD 400-950 (Lamb, 1939: 10; Hill and Kivett, 1940: 211)
Bakenhus ossuary, 25PT4, Sec. 15, T19N, R1E; age—AD 400-950 (Lamb, 1939b: 1)
Feye Site, 25PT9, 5.5 mi. S Creston; age—AD 0-1000 (Kivett, 1952: 43-47)
Hill-Rupp [Monroe] Site, 25PT13, Sec. 25, T18N, R2W, 1.5 mi. N Monroe; age—AD 1650-1750 (Metcalf, 1941a: 13)

Polk Co.
Clarks Site, 25PK1, Sec. 17, T14N, R4W; age—AD 1820-1849 (Blackman, 1907: 358)

Red Willow Co.
25RW22, 3 mi. N, 6.25 mi. W Indianola; age—AD 1200 (Grange 1980: 95-99; Mick, 1983: 67-68)
25RW23, 3 mi. N, 6.25 mi. W Indianola; age—AD 1700-1750 (Grange, 1980:143)
Coon Creek Site, 25RW24, west edge Indianola; age—AD 1000-1350 (Grange, 1980: 101-103)
25RW25, 3.25 mi. N, 6.25 mi. W Indianola; age—AD 1000-1350 (Grange, 1980: 143)
Doyle Site, 25RW28, 9.5 mi. N, 0.75 mi. W McCook; age—BC 60-AD 680 (Grange, 1980: 105-126)
Gillen Pits, 25RW101, Sec. 28, T3N, R30W, 4.5 mi. W McCook; age—Late Pleistocene-early Holocene (Corner, 1977: 77-79)
Davidson Pits, 25RW102, Sec. 11, T3N, R27W, 1 mi. W Bartley; age—Late Pleistocene-early Holocene (Corner, 1977: 77-79)
Redwillow 1, 6 mi. NW McCook; age—unknown (Wedel, 1934: 171)

Richardson Co.
Leary Village Site, 25RH1, lower Nemaha River, 4 mi. SE Rulo; age—AD 800-1500 (Hill and Wedel, 1936: 4-8; Blakeslee and Caldwell, 1979: 1-22)
Rulo Southeast, 25RH69, 0.4 km S Rulo; age—AD 1037-1396 (Nelson, 2006: 165)
Rulo Southeast, 25RH70, 0.6 km S Rulo; age—AD 1041-1284 (Nelson, 2006: 170, 173)

Sarpy Co.
Childs Point District, 25SY0, ≈ 1.5 mi. NW Bellevue; age—AD 1050-1425 (Gilder, 1909: 56-60)
Fremont 1 [Farnsworth Site], 25SY1, margin of uplands overlooking bottom lands above Elkhorn and Platte rivers, 3 mi. N, 2 mi. W Gretna; age—AD 1300-1400 (Hill and Cooper, 1936c: 271)
Fontenelle Forest Site, 25SY3, 1 mi. N Bellevue; age—AD 1100-1450 (Hill and Cooper, 1937: 298-304)
Gates Site, 25SY5, probably near modern Capehart Road and 36th Street, 4 mi. W Bellevue; age—AD 1000-1400 (Strong, 1935:143-145)
Lucien Fontenelle's Post, 25SY26, Bellevue; age—AD 1822-1842 (Bozell et al., 1990: 1)
Patterson Site, 25SY31, ≈ 7 mi. S Gretna; age—AD 1000-1300 (Bozell and Ludwickson, 1999)
unnamed site, near north side of Bellevue; age—unknown (Gilder, 1909: 78)
unnamed site, near Meadow; age—unknown (Brooking, 1921: 37)

Scotts Bluff Co.
Signal Butte Site, 25SF1, Signal Butte, 11.5 mi. S Morrill; age—BC 3000-AD 500 (Koch, 2000; Strong, 1935: 224-227)
Scottsbluff Bison Quarry, 25SF2, 16 mi. W, 3 mi. S Scottsbluff; age—BC 8250 (Schultz and Eiseley, 1935: 307; Hill, 2008: 91; Knudson, 2013: 294-295)
Gering Burial Site, 25SF10, Sec. 14, T21N, R54W, 1.5 mi. NW Melbeta; age—AD 600-800 (Oothoudt, 1976: 19-23, 166-167)
25SF11, Scotts Bluff National Monument, summit Scotts Bluff, 0.45 mi. N, 0.3 mi. E Visitor Center; age—1000-1450 (Bozell, 1994b: 49-63; Clark, 1994b: 11)
25SF39, Scotts Bluff National Monument, 0.2 mi. S, 0.2 mi. W Visitor Center; age—unknown (Bozell, 1994b: 49-63; Clark, 1994b: 14)
25SF50, Scotts Bluff National Monument, 0.9 mi. N, 0.4 mi. E Visitor Center; age—AD 1022-1635 (Griffin, 1991: 1-55)
Scott's Spring, 25SF51, Scotts Bluff National Monument, 0.2 mi. N, 0.4 mi. E Visitor Center; age—AD 657-1277 (Bozell, 1994b: 49-63; Clark, 1994b: 18-23)
25SF53, 3.25 mi. S, 0.75 mi. E Lyman; age—BC 6000-0 (Koch and Miller, 1996: 1-7, 12-15, 108)
25SF65, Scotts Bluff National Monument, 0.7 mi. N, 0.1 mi. E Visitor Center; age—AD 1311-1624 (Bozell, 1994b: 49-63; Clark, 1994b: 26-28)
25SF76, Scotts Bluff National Monument, 0.7 mi. N, 0.1 mi. W Visitor Center; age—AD 1100-1500 (Bozell, 1994b: 49-63; Clark, 1994b: 29-32)

Sherman Co.

Bill Packer Site, 25SM9, along Davis Creek in extreme northeastern corner of county; age—AD 980-1050 (Graham et al., 1987: 240; Bozell and Rogers, 1989: 3-5; Souders, 1994: 45, 53)
Sherman Feeder Canal, 25SM203, Sec. 5, T16N, R15W; age—unknown (Cumming and Stephenson, 1953: 6)
Sweetwater Site, Sec. 35, T13N, R15W [41º02.5'N, 99º01'W], 0.5 mi. NW Sweetwater [Buffalo Co.]; age—AD 1250-1450 (Champe, 1936: 253-257)

Sioux Co.
25SX26, Agate Fossil Beds National Monument, 0.4 mi. N, 0.6 mi. E Visitor Center; age—unknown (Clark, 1993: 9, 44)
Cedar Canyon #1-#3, #7, 25SX103, Sec. 16, T33N, R53W; age—unknown (Meston, 1976: 21-23)
Hudson-Meng Site, 25SX115, 23 mi. NW Crawford [in Dawes Co.], 42°44'22"N, 103°36'10"W; age—BC 7820 (Agenbroad, 1977, 1978)
Whitehead West #2, 25SX116, Sec. 13, T33N, R54; age—circa 1700 (Meston, 1976: 18-19)
Sand Creek 3-4, 25SX141, Sec. 28, T33N, R 53W; age—unknown (Meston, 1976: 37)
Sand Creek 3-5, 25SX142, Sec. 28, T33N, R53W; age—unknown (Meston, 1976: 38)
Milo Saxton, 25SX146, Sec. 28, T33N, R53W; age—AD 0-1000 (Meston, 1976: 35-36)
25SX153, Agate Fossil Beds National Monument, 0.5 mi. W Visitor Center; age—unknown (Clark, 1993: 10-11, 44)
25SX157, Agate Fossil Beds National Monument, 19.25 mi. S, 5.75 mi. E Harrison; age—BC 6000-1000 (Bozell, 1994a: 46-49, 61-63; Clark, 1994a: 1)
25SX163, Agate Fossil Beds National Monument, 19.25 mi. S, 7.25 mi. E Harrison; age—BC 1000-AD 500 (Bozell, 1994a: 49, 61-63; Butler, 1994: 29)
25SX268, Agate Fossil Beds National Monument; age—unknown (Galindo, 1997: 120-126.
25SX471, Agate Fossil Beds National Monument, 0.4 mi. E Agate; age—unknown (Clark, 1993: 18, 44)
25SX472, Agate Fossil Beds National Monument, 0.65 mi. E Agate; age—unknown (Clark, 1993: 18, 44)
25SX475, Agate Fossil Beds National Monument, 1.5 mi. E Agate; age—unknown (Clark, 1993: 19-20, 44)
25SX476, Agate Fossil Beds National Monument, 1.7 mi. E Agate; age—AD 900 (Clark, 1993: 20, 44)
25SX486, Agate Fossil Beds National Monument, 19.25 mi. S, 5.25 mi. E Harrison; age—unknown (Bozell, 1994a: 50, 61-63; Clark, 1994a: 11)
25SX487, Agate Fossil Beds National Monument, 0.1 mi. N, 0.15 mi. E Visitor Center; age—unknown (Clark, 1993: 23-24, 44)

Stanton Co.
Stanton Site, 25ST1, 1 mi. E Stanton; age—AD 1200-1500 and AD1800-1833 (mixed site) (Gunnerson, n.d.)

Thayer Co.
Rose Creek, Sec. 29, T1N, R1W; age—AD 1700-1850 (Lamb, 1932: 169-172)

Thurston Co.
Gerald Rose Site, 25TS10, Sec. 17, T26N, R8E, vicinity of Winnebago; age—AD 1000-1500 (Gant, 1966: 5; Carlson, 1973: 52-54)
Omaha Creek Site, 25TS11, Sec. 27, T26N, R8E; age—AD 1000-1500 (Gant, 1966: 6)

Valley Co.
Mira Creek Site, 25VY1, 6 mi. W, 0.75 mi. N North Loup; age—AD 0-1000 (Hill and Kivett, 1940: 183)

Washington Co.
Fort Atkinson Site, 25WN9, 0.5 mi. E Fort Calhoun; age—AD 1819-1827 (Carlson, 1979: 1-12)
Site W1, on Douglas-Washington Co. line in extreme southeastern corner of county; age—AD 0-1000 (Cooper, 1939: 107-111)

Webster Co.
Hill Site [Pike Pawnee Site], 25WT1, Sec. 13, T1N, R10W, 2 mi. S, 7 mi. E Red Cloud; age—AD 1700-1850 (Wedel, 1936: 34; Metcalf, 1941b: 3)

Shipman Site, 25WT7, south side of Republican River, between Red Cloud and Guide Rock; age—post AD 1350 (Mick, 1983: 67; Graham et al., 1987: 240)
Superior 2, 3 mi. W Guide Rock, age—unknown (Wedel, 1934: 143)
Red Cloud 1, Red Cloud; age—unknown (Wedel, 1934:148)

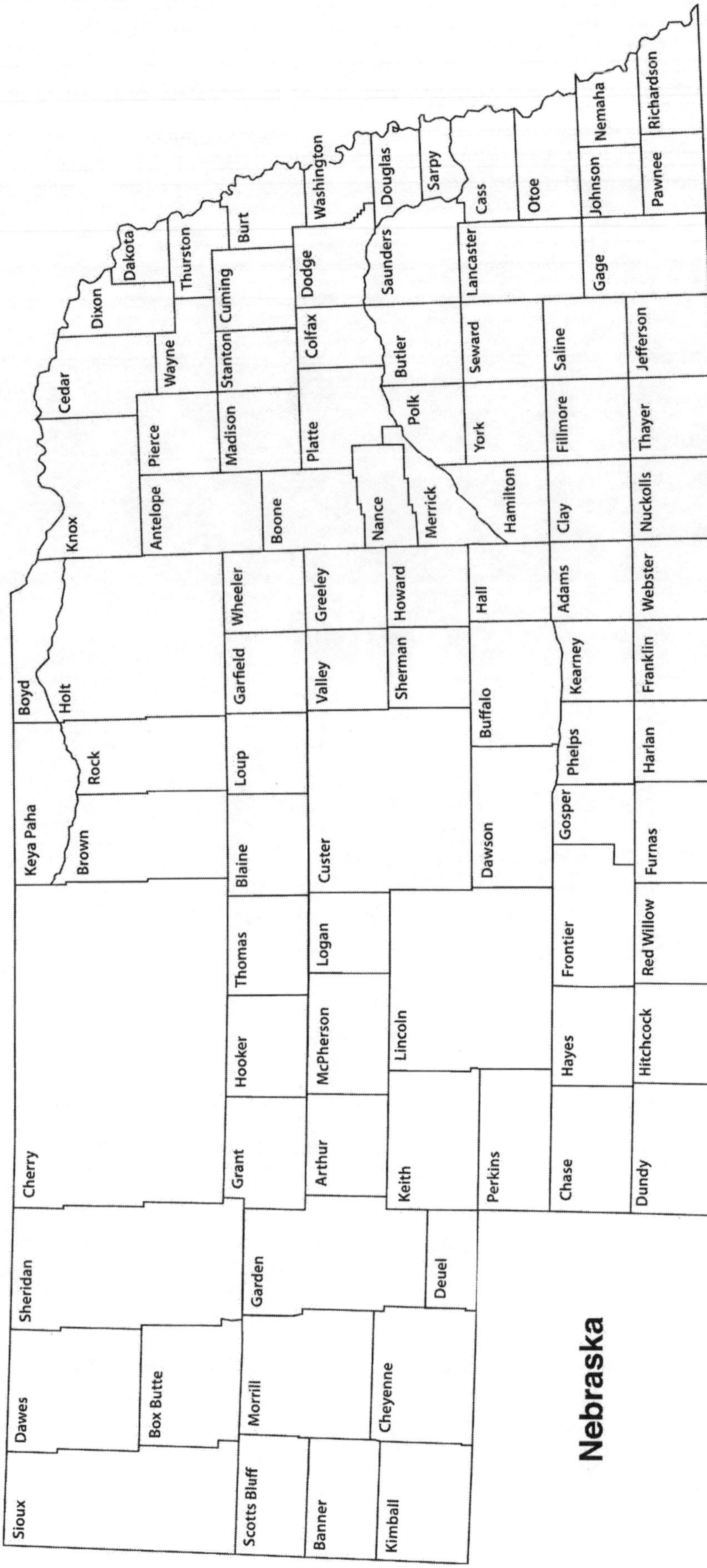
Nebraska
Sioux
Dawes
Sheridan
Cherry
Keya Paha
Boyd
Knox
Cedar
Dixon
Dakota
Scotts Bluff
Box Butte
Brown
Rock
Holt
Antelope
Pierce
Wayne
Thurston
Banner
Morrill
Garden
Grant
Hooker
Thomas
Blaine
Loup
Garfield
Wheeler
Boone
Madison
Stanton
Cuming
Burt
Kimball
Cheyenne
Deuel
Arthur
McPherson
Logan
Custer
Valley
Greeley
Nance
Platte
Colfax
Dodge
Washington
Keith
Lincoln
Sherman
Howard
Merrick
Polk
Butler
Saunders
Douglas
Sarpy
Perkins
Dawson
Buffalo
Hall
Hamilton
York
Seward
Lancaster
Cass
Otoe
Chase
Hayes
Frontier
Gosper
Phelps
Kearney
Adams
Clay
Fillmore
Saline
Gage
Johnson
Nemaha
Dundy
Hitchcock
Red Willow
Furnas
Harlan
Franklin
Webster
Nuckolls
Thayer
Jefferson
Pawnee
Richardson

Made in the USA
Monee, IL
04 November 2026

67383415R00055